Plato's Revenge

Plato's Revenge
Politics in the Age of Ecology

William Ophuls

The MIT Press
Cambridge, Massachusetts
London, England

First MIT Press paperback edition, 2013

© 2011 Massachusetts Institute of Technology

All rights reserved. No part of this book may be reproduced in any form by any electronic or mechanical means (including photocopying, recording, or information storage and retrieval) without permission in writing from the publisher.

For information about special quantity discounts, please email special_sales@mitpress.mit.edu.

This book was set in Sabon by Toppan Best-set Premedia Limited. Printed and bound in the United States of America.

Library of Congress Cataloging-in-Publication Data

Ophuls, William, 1934–
Plato's revenge : politics in the age of ecology / William Ophuls.
  p. cm
Includes bibliographical references and index.
ISBN 978-0-262-01590-5 (hardcover : alk. paper)—978-0-262-52528-2 (pb.)
1. Environmental policy. 2. Political science—Philosophy. I. Title.
GE170.O65 2011
320.01′5—dc22
                                                         2010053624

10 9 8 7 6 5 4 3 2

For Harriet, Auden, and Cyrus

Caught in the relaxing interval between one moral code and the next, an unmoored generation surrenders itself to luxury, corruption, and a restless disorder of family and morals.
—Will and Ariel Durant, *The Lessons of History*

Contents

Preface   ix

Prologue: The Five Great Ills   1

The Necessity of Natural Law

1   Law and Virtue   13

The Sources of Natural Law

2   Ecology   25
3   Physics   45
4   Psychology   69

The Politics of Consciousness

5   *Paideia*   97
6   *Politeia*   129
7   A More Experienced and Wiser Savage   167

Epilogue: True Liberation   195

Bibliographic Note   199
Notes   209
List of Sources   231
Index   247

Preface

This book completes the task I set myself many years ago—to find a humane and effective political response to the challenge of ecological scarcity. The challenge arises from an ensemble of interlocking biological, geological, and physical limits that now threatens the welfare and possibly the existence of industrial civilization.

I began in 1977 with *Ecology and the Politics of Scarcity* by arguing that modern political economies devoted to "economic development" were on a collision course with the laws of ecology, which forbid perpetual growth. The entire set of modern ideas, institutions, and practices predicated on abundance would have to be replaced by ones grounded on scarcity, which was about to reemerge with unexpected force. Whether they liked it or not, human beings would soon be obliged to renounce continual material growth and instead devise a steady-state political economy that would make it possible for humanity to live in a long-term, harmonious balance with nature.

Because the conquest of scarcity by technological means is a modern article of faith, this ecological argument was mostly dismissed out of hand on the ground that market forces and human ingenuity would ever and always vanquish material constraints. To the extent that the argument was accepted as real, it was mostly misunderstood, and the political response—first to frame the problem in terms of environmental concern rather than ecological peril and then to try to mitigate waste and pollution while ignoring the antiecological dynamic that created them—was therefore superficial. In short, it was business as usual, with some minor modifications intended to treat the most glaring symptoms of ecological illness.

This failure to grasp that the root of the disease is not defective public *policies* but a defective public *philosophy* motivated me to resume the discussion in 1997 with *Requiem for Modern Politics*. In that work, I argued that the modern political paradigm—that is, the body of political concepts and beliefs inherited from Thomas Hobbes and his successors—was bound for self-destruction even before the emergence of ecological scarcity. That paradigm is no longer intellectually tenable or practically viable because any polity that abandons virtue and rejects community necessarily becomes the author of its own demise. The tendencies toward moral decay, social breakdown, economic excess, and administrative despotism that are evident everywhere in the so-called developed world testify to the need for a new public philosophy—on political as well as ecological grounds.

This book attempts to sketch the basic outline of such a philosophy—a natural law theory of politics grounded in ecology, physics, and psychology. In doing so, I make explicit the basic principles of ecological polity that were implicit in my previous work and add new material to make the theory more robust.

I start from the radical premise that "sustainability" as usually understood is an oxymoron. Industrial man has used the found wealth of the New World and the stocks of fossil hydrocarbons to create an antiecological *Titanic*. Making the deck chairs recyclable, feeding the boilers with biofuels, installing hybrid winches and windlasses, and every other effort to "green" the *Titanic* will ultimately fail. In the end, the ship is doomed by the laws of thermodynamics and by implacable biological and geological limits that are already beginning to bite. We shall soon be obliged to trade in the *Titanic* for a schooner—in other words, a postindustrial future that, however technologically sophisticated, resembles the preindustrial past in many important respects. This book attempts to envision the politics of that smaller, simpler, humbler vessel.

From this radical premise, which will not be easily accepted, follows a return to first principles. Rather than joust with contemporary specialists in environmental affairs, I therefore decided from the outset to rely on the time-tested classical authors who have eloquently and cogently grappled with the core issues of politics. I found Plato, Jean-Jacques Rousseau, and the rest more illuminating and pertinent than other writers to achieving my aim—namely, encouraging readers to question our most basic social, economic, political, and even moral assumptions as the first step toward imagining a truly ecological future.

To put it another way, I have written a provocative essay, not a scholarly treatise. The measure of my success will be its ability to challenge convention and to move the conversation beyond a misguided search for so-called sustainability and toward a deeper examination of our basic principles. Ecological scarcity is not a problem that can be solved within the old framework but a predicament or dilemma that can be resolved only by a new way of thinking. The ultimate outcome of this new philosophy will be a new political order.

My essay does not describe this new order. Nor do I prescribe a particular form of government or its precise powers. Instead, I focus on the epistemological, ontological, and ethical basis of politics and then suggest where this might lead. Nevertheless, it should be clear where my sympathies lie—with a fundamentally limited, Jeffersonian, republican form of government. This is not out of mere predilection but because that seems to me to be what ecological reality demands.

But we must keep an open mind. The experience of living in different cultures has taught me that no one form of government fits all peoples and all circumstances. In the end, there is wisdom in Alexander Pope's famous couplet:

For Forms of Government let fools contest;
Whate'er is best administered is best.

Since good administration necessarily varies according to the prevailing conditions within a society, I have tried to frame my theory so that it can be the basis for many different kinds of political arrangements, from tribal monarchy to direct democracy.

Some may object that a radical change in public philosophy is hardly a practical or feasible solution—as if it were somehow illegitimate to propose answers to our problems that do not accord with received ideas or that cannot be implemented by existing institutions. But if our problems have been created by a certain way of thinking, then the only real solution is to adopt a new way of thinking and not to devise clever political or economic mousetraps based on the old one. As Albert Einstein is supposed to have said, "No problem can be solved from the same level of consciousness that created it." And once adopted, the new level of consciousness will almost automatically generate the requisite practical measures. Why is it taboo to propose political change when we complaisantly permit massive, unlegislated technological changes that have the effect of overturning the social order? The current American political regime is not sacrosanct. If the founders could see how the Constitution that they framed with such prudence has been subverted by their political progeny, they would be appalled. The only genuine solution to our predicament is a new political philosophy, however impractical, unfeasible, or even heretical it might seem to adherents of the old one.

For some, political philosophy is irrelevant for all practical purposes because technique and finance, not poetry, now legislate for humankind. This makes our governing ideas mere resultants or rationales, not causes. But John Maynard Keynes argued to the contrary that practical men of affairs are in reality the intellectual slaves of defunct scribblers. In our case, we are the slaves of Thomas Hobbes. Despite his lament at the end of book 2 of *Leviathan* that his philosophical labor was as "useless" as Plato's *Republic*, Hobbes's ideas, as revised and elaborated by John Locke and Adam Smith, became the template of modern life—that is, life seemingly determined by technique and finance. In other words, the economic and technocratic juggernaut driving us toward an increasingly chaotic and dismal future is but the physical manifestation of Hobbes's mostly unacknowledged philosophy. Until we invent and implement a better one that is inspired by a vision of a more satisfying and genuinely sustainable future, nothing can change for the better. In the end, not only do ideas matter, but they may be all that matter. As Keynes said, "the world is ruled by little else."[1] The process is inevitably dialectical: when ideas are given concrete form, that form then affects our way of thinking. To adapt Winston Churchill's tribute to the power of architecture, "We shape our institutions, and afterward our institutions shape us."

It is customary for authors to thank their benefactors. My gratitude for so many blessings—among them the people, books, and experiences that made me who I am and that shaped my way of thinking—knows no bounds. Regarding my more immediate debts, this is an essay, not a monograph, so I have cited only major influences and borrowings. Let me therefore acknowledge the many others who have also struggled to come to grips with the ecological problematique in all its manifold aspects but whose works are not cited here. They have prepared the ground for my more radical venture. May this work in some small way requite these many debts.

My first book was dedicated "To the posterity that has never done anything for me" because I found the selfishness and casual destructiveness of our consumptive way of life morally obnoxious as well as ecologically blind. This one is also dedicated to posterity in the concrete form of my grandchildren. Let them be the stand-ins for the future generations who will have to suffer the painful legacy of past ecological malfeasance and who will have to discover an entirely new way of being civilized on a crowded, finite planet. To all of them, I extend my condolences for the impoverished and poisoned earth they stand to inherit. I hope that they will succeed where we have failed in making a civilization worthy of the name.

Prologue: The Five Great Ills

Opinions alter, manners change, creeds rise and fall, but the moral laws are written on the tablets of eternity.
—James Anthony Froude[1]

Since its origin, civilization has been marked by five great ills—ecological exploitation, military aggression, economic inequality, political oppression, and spiritual malaise. Our precivilized ancestors, however, should not be idealized or romanticized. Their lives were arduous, their habits often squalid, and their mores sometimes savage. Violent death was common. Nor were they ecological angels: before they learned to live in balance, they exterminated fauna and ravaged flora.[2] Yet despite the charges that could be levied against them, they eventually evolved ways of living in harmony with the earth and with each other. Above all, they enjoyed what the pioneering anthropologist Lewis Henry Morgan called "the liberty, equality, and fraternity of the ancient Gentes," the memory of which still lingers in the collective consciousness of the human race.[3] Finally, they possessed a natural religion without doctrines, priests, or churches—the famous *participation mystique*, which connected them to the cosmos and gave meaning to their lives.

From this perspective, the rise of civilization constitutes a Faustian bargain or even a tragic fall from primal grace. When human beings abandoned the ecological niche in which they had evolved, they left a state of natural plenitude, however rough, for a life of toil in field and mine. They became more numerous and prosperous but less healthy.[4] The technological means that they used to enrich themselves also harmed nature and turned war from a blood sport into a vehicle for conquest or extermination. Liberty was replaced by authority, equality by hierarchy, and fraternity by disunity. The many, who had once lived in small bands as kinsmen and equals, became subject to the few—to the emperors, kings, and tyrants who expropriated the wealth they produced. Natural religion gave way to organized religion, whose priests, rites, and doctrines served mostly the oppressors' interests, even as they gave some solace to the estranged denizens of the ancient cities. In short, the indisputable advantages of civilization were purchased at a high price.[5]

Much of this was apparent to the philosophers and statesmen who created the modern world, but their diagnosis of the disease—and therefore their proposed treatment—was flawed. They sought to cure two of the five great ills (economic inequality and political oppression) by intensifying two others—ecological exploitation and military aggression. As a result, the modern age is marked by the ethos of the conquistador. Scientists master nature in their laboratories so that engineers can build arsenals and factories, manufacturers can make arms and goods, and soldiers and merchants can dominate the lands and markets of the world.

These thinkers were driven by a quest for power—for dominion over nature, which would foster dominion over the world. But as Lord Acton famously said, power corrupts, and the more absolute the power, the worse the corruption. Indeed, power seems to drive men and women mad, with hubris being the worst symptom of the disease.

The response of the Enlightenment *philosophes* to the fifth great ill was equally problematic. They set about liberating men and women from clerical religion because they detested the venality, inquisitorial zeal, and reactionary politics of the established church, and they succeeded all too well in crushing Voltaire's *infâme*. When the babe of morality was thrown out with the bathwater of superstition, the consequence was a process of demoralization that began slowly but has now become a rout.

This demoralization has three aspects—the corruption of morals and mores, the undermining of morale, and the spreading of confusion—and has resulted in the loss of almost all sense of honor, duty, and responsibility. Solidarity, too, has eroded, as individuals and groups engage in a winner-take-all struggle for power and wealth. However glutted with goods people in rich countries may be, they feel that they are subject to a vast, impersonal, out-of-control system that gives them the vote, that mostly abides by juridical rules, but that denies them real liberty and equality. Fraternity is not even an issue. Last but not least, because God is dead and only instrumental reason counts, all authority and orientation have been overthrown—so men and women have lost not only their intellectual and spiritual bearings but even the means by which to take them.

The five great ills of civilization therefore have become evils that threaten the continued existence of human society. Ecological exploitation has degenerated into the systematic and ruthless abuse of nature, causing an accelerated degradation and depletion of our natural milieu. We ourselves have begun to suffer certain inconveniences, and our grandchildren stand to inherit a poisoned and impoverished planet. Indeed, as the age of petroleum draws to a close, the material basis for an advanced technological culture capable of supporting billions of people in sprawling megacities is by no means assured.

Similarly, military aggression has escalated into potential holocaust, as weapons of mass destruction are ever more widely disseminated. And wars are no longer fought by brave warriors and wily generals who meet face to face on a battlefield but by military bureaucrats and technicians who risk nothing as they rain electronic death on remotely seen enemies—or unarmed innocents.

In the same way, our economic system has vastly amplified the scope and scale of economic inequality. Despite a general rise in material well-being, wealth is radically maldistributed, and billions of people continue to live in destitution and misery. In addition, the rich command resources unimaginable to ancient kings, so the rod by which deprivation is measured has grown enormously.

Nor has political oppression vanished. Even in states where the principle of liberty is well established, the burden of bureaucratic regulation becomes ever more minute, all encompassing, and suffocating. Traditional liberties are being eroded in the name of expediency in efforts to defend national security and fight terrorism, crime, drugs, and tax evasion. A sphere of privacy hardly exists anymore. Meanwhile, democracy is mostly a sham: either money rules, or remote policy elites in cahoots with powerful economic interests make all the important decisions.

Lastly, spiritual malaise is pandemic. As a result, demoralized individuals must struggle to keep their psychic footing. Many resort to diseased methods of coping, not only physical addiction to drugs, alcohol, and tobacco but also psychological addiction to eating, entertainment, gambling, pornography, sex, shopping, and sports. Many simply cannot cope. The armies of social workers and psychotherapists may help a handful of individuals, but they can do little to save society, which becomes fertile ground for every form of mania.

This demoralization was never intended by the thinkers who created the modern world. Believing as they did (and not without reason) that organized religion was an almost unmitigated evil, they sought to liberate us from religious politics—from the interference of an established church in the public affairs of the state and the private affairs of the individual. Thanks to their efforts, we in the West are no longer subject to clerical oppression or to a despotic form of spirituality, for which we must be eternally grateful.

But we have paid a steep price for this liberation. Indeed, far from creating a rational utopia, banishing superstition and exalting reason have created a spiritual void that has been filled by absurd and dangerous political, social, and economic ideologies that have often proven to be as pathological in their historical consequences as the dogmatic religions of old.

In retrospect, it may seem surprising that the *philosophes* had so few qualms about crushing the established church, one of the pillars of the existing social order. But they believed that traditional religion was dispensable precisely because they were certain that human reason, once liberated from theology, would soon discover the moral order implicit within the cosmos—an order to which men and women, being reasonable beings, would naturally and willingly accede.

That did not happen. The secularization promoted by the Enlightenment took on a logic and momentum of its own. Rationalism displaced reason, so the only permissible natural laws were mechanical, not moral. Human beings also turned out to be far less reasonable and much more irrational than these thinkers assumed.

The triumph of secularism has had consequences that are devastating in the political sphere. A purely rational and material politics—a politics without a moral code or a vision of the good life or a sense of the sacred—is a contradiction in terms. As Aristotle pointed out, no polity can long exist as "a mere alliance" of self-interested individuals.[6] What makes a political community cohere is what Aristotle called "a rule of life"—that is, a shared ethos.[7]

But the rule of life of modern politics is that we shall have no positive rules, only negative ones that keep us from harming others but that otherwise leave us at liberty. The citizens themselves must sustain community through social institutions—churches, schools, voluntary associations, informal networks—that inculcate a shared ethos and foster a sense of common destiny. In other words, the indispensable linchpin of the modern state is civil society, for it alone supplies the cohesion that a liberal polity lacks.

Unfortunately, the process of demoralization described above has effectively destroyed the morals, mores, and morale of civil society. As a result, polity today is more and more a mere alliance of self-interested individuals who pursue their own private ends and who accept only minimal restraints on their actions. Liberty has become license, and the social basis of the modern, liberal state has eroded away.

In effect, the project of modern politics has failed. When Hobbes took the radical step of severing politics from virtue and founding the polity on the self-interested individual, he started a movement that liberated men and women from subservience to king and bishop, but he also set in motion a vicious circle of moral decay that has all but overwhelmed civil society. The legal and bureaucratic machinery of government has grown larger and more oppressive in a mostly vain attempt to make up for social decline. We are being driven toward an administrative despotism that extinguishes both liberty and privacy because it is the most expedient way to deal with the moral breakdown caused by our basic political principles.

It is bad enough that a secular and rational politics has destroyed its own foundation and now seems bent on creating a Leviathan. What is even more dangerous is that casting men and women loose from their traditional cultural and religious moorings leaves them adrift in a meaningless cosmos, lacking clear metaphysical or practical answers to the basic problems of life.

The resulting spiritual vertigo is responsible for much of the social and personal dysfunction mentioned above and also for the calamitous history of the twentieth century. Only a few artists, philosophers, and free spirits thrive on the radical openness of cultural nihilism. The average person hates it, and if people do not get satisfactory answers to the questions of life from their inherited culture, then they will seek them elsewhere. This explains the popular appeal of the fanatical ideologies that drenched the last century in blood (and of the religious fundamentalism that now threatens to do the same in this one).

In reality, the Enlightenment did not so much abolish religion as redirect the spiritual drive of the Judeo-Christian tradition toward worldly ends. We moderns are just as religious as our premodern ancestors, but we have chosen to worship two savage gods—Moloch and Mammon. Those who worship Moloch turn politics into a perverted religion. They try to fill the void caused by cultural nihilism with eschatological secular creeds dedicated to achieving a utopian ideal of social perfection. Those who worship Mammon turn politics into a religion of the self. They try to fill the void by glutting themselves with pleasure, exalting their own self-gratification into a moral principle and exploiting the state for selfish ends. These are both false gods. Neither ideology nor self-indulgence can satisfy the spiritual needs of human beings or make them truly happy, and both tend toward destruction.

Our secular, rational, amoral way of life is failing. Our cultural myth to the contrary notwithstanding, this way of life represents not a final progressive advance of civilization to "the end of history" but an intensification of civilization's inherent flaws that can end only in tragedy. We must reinvent civilization so that it once again rests on a moral foundation by discovering a new "rule of life" that moderates, rather than magnifies, the five great ills. And we now have the means to do so. The epistemological and ontological revolution of the twentieth century that produced systems ecology, particle physics, and depth psychology reveals a moral order that is immanent within the scientific description of the universe. From this order—"written on the tablets of eternity"—we can derive principles that could form the basis for humane and prudent governance. In other words, we have rediscovered the kind of natural law that the *philosophes* envisioned. We now understand, better than our Enlightenment ancestors, the means by which we can actualize these principles without resurrecting the evils of organized religion.

In this book, I begin by examining the role played by law in human society before showing that ecology, physics, and psychology all agree in pointing us toward a politics of consciousness dedicated to expanding human awareness rather than extending human dominion. Unless the means of civilization are soon directed to an end that is higher than the endless accumulation of wealth and power, then the very enterprise of civilization itself, not just our particular form of it, may not long survive.

The Necessity of Natural Law

1
Law and Virtue

The more corrupt the state, the more numerous the laws.
—Tacitus[1]

Modern life is fundamentally lawless. We have an abundance or even a surfeit of man-made laws, but legislated morality is an inadequate and potentially dangerous expedient made necessary by the absence of a basic moral order, the fundamental ground of any society. Without that ground, piling law upon law will hasten, rather than forestall, the onset of social and political breakdown.

Legislation is no substitute for morality. For Aristotle, "The best laws, though sanctioned by every citizen of the state, will be of no avail unless the young are trained by habit and education in the spirit of the constitution."[2] For Hippolyte Taine, "the aim of every society must be a state of affairs in which every man is his own constable, until at last none other is required."[3] Mores, not laws, make us law-abiding and public-spirited.

To put it the other way around, in the absence of an inner disposition to behave morally, people will inevitably find ways to avoid, evade, subvert, delay, or otherwise frustrate the operation of laws. This precipitates a legal arms race. Self-seeking individuals, unrestrained by virtue, seize opportunities to bend the law to their own selfish ends, and this behavior requires yet more legislation to close the loopholes, and so on ad infinitum. The upshot is a labyrinth of laws that grows ever more convoluted and corrupt.

To spell out the implied syllogism of Tacitus,

- In a healthy state, laws are few, simple, and general because the people are moral, law-abiding, and public-spirited, which makes them easy to govern.
- In a sick state, the laws are many, complex, and minute because the people are amoral, conniving, and self-seeking, which makes them hard to govern.
- Ergo, the more numerous the laws, the more corrupt the state, and vice versa.

By this standard, the United States is hopelessly corrupt. In fact, it may be the most law-ridden society that has ever existed. The volume and complexity of statutes and the rapidity with which they are amended make a mockery of the legal fiction that "ignorance of the law is no excuse."[4] Even full-time specialists find the labyrinth daunting, and the bureaucrats who inflict the laws on the public repeatedly err in their interpretation of them.

The laws are not only increasingly numerous, complex, and all encompassing, but they are also more draconian and even tyrannical. Indeed, said Edmund Burke, "Bad laws are the worst sort of tyranny."[5] For example, mandatory sentencing means that judges cannot temper justice with mercy or common sense, so prisons are packed with petty criminals, candidates for rehabilitation, as well as violent felons. The talk of an American gulag is not idle: legislating morality has real social costs.⁶ Nor, thanks to civil forfeiture and other legal bludgeons handed to prosecutors in recent years, can defendants always expect to mount an adequate defense. One of those bludgeons, the Racketeer Influenced and Corrupt Organizations Act (RICO), effectively abolishes the presumption of innocence on which the criminal justice system supposedly rests.

We also seem to be slouching toward a police state in which *raison d'état* trumps civil rights. Exploiting popular fear of criminals and terrorists, successive presidential administrations have concentrated more and more coercive power in the hands of the state—a power that operates behind a shield of secrecy to strip away long-established rights to privacy and liberty. In a development that evokes memories of life behind the Iron Curtain, bankers, therapists, pharmacists, teachers, and other civilians are now legally compelled to spy on their fellow citizens.⁷

As indicated in the prologue, the law-ridden and corrupt modern state, of which the United States is a preeminent exemplar, results from a demoralization that is an inescapable consequence of our basic political principles. And the process is all but irreversible. As with Gresham's law in economics, bad values drive out good, so the moral currency is continuously debased.

Once demoralization is well advanced, reform efforts exacerbate the problem. Without a general consensus, the attempt by some to impose morality by law on others embroils society in perpetual warfare over issues such as crime, drugs, abortion, and schooling. Legal substitutes for morality are therefore a symptom of the disease rather than a cure for it. They do not arrest moral decay but advance the corruption of the state. When laws are no longer felt to be general principles of justice but instead are seen to be the product of organized selfishness, factional strife, or moralistic meddling, then all respect for law is lost—and even the legitimacy of the state is called into question.

But how can our basic political principles foment demoralization and the dangerous consequences described above? Is not liberal polity the final answer to the riddle of politics—or at least a better answer than all the others? Despite the banner of progress under which they march, liberal polities are self-destructive. Both in theory and practice, they depend on reason and self-restraint—that is, on citizens who know the difference between liberty and license and who govern themselves accordingly. But the intrinsic amorality of liberalism first erodes, then corrodes, and finally dissolves these faculties.

To make a long story short, all modern polity is rooted in Hobbes's rejection of the classical conception of the polity—namely, that the state has a duty to make men and women virtuous in accordance with some communal ideal. Instead, said Hobbes, let individuals follow their own ideals and pursue their own ends with the state acting simply as a referee to prevent injury or harm to others. Hence, the function of the state is purely instrumental: it keeps the peace and relegates morality to the private sphere.

Unfortunately, by making politics instrumental rather than normative, Hobbes and his followers set up a vicious circle leading to demoralization. If individuals are left to their own moral devices with nothing but rationality to guide them, said Will Durant, there can be no other outcome:

The brilliant enfranchisement of the mind sapped the supernatural sanctions of morality, and no others were found to effectually replace them. The result was such a repudiation of inhibitions, such a release of impulse and desire, so gay a luxuriance of immorality as history had not known since the Sophists shattered the myths, freed the mind and loosened the morals of ancient Greece.[8]

Mores are a matter not of rational calculation but of heartfelt conviction. Reason may (as Hobbes and his liberal followers argued) instruct us in virtue, but this is likely to be effective only for philosophers. The rest of us need stronger medicine. Without such medicine, the sentiment that keeps individuals law-abiding even in the absence of positive law is fated to grow ever weaker as reason succumbs to passion. As Durant suggests, far from inculcating moral restraint, a reason that is excessively rational becomes part of the problem. Rationality may begin benignly by liberating us from superstition, but after disposing of myth and religion along the way, it ends by ruthlessly deconstructing every form of meaning or authority.

Amorality and nihilism were not problems at the origin of the modern era because society was for many years sustained by the virtues and beliefs inherited from the premodern era. As this lode of fossil virtue and belief was eroded away, however, individuals became increasing self-seeking, amoral, and even immoral.

The moral calculus of liberal politics was succinctly stated by the Marquis de Sade, who adumbrated with his own depraved conduct the fateful consequences of allowing mores to be a matter of private choice. Writing in 1797, de Sade noted that to adopt self-interest as "the single rule for defining just and unjust" was to make morality a fiction: "There is no God in this world, neither is there virtue, neither is there justice; there is nothing good, useful, or necessary but our passions, nothing merits to be respected but their effects."[9]

To avoid anarchy, the decline in inner lawfulness that follows demoralization necessarily demands an increase in outer compulsion. The modern state has been obliged to step in to replace a civil society whose vigor has been sapped by moral entropy. In short, just as Hobbes himself maintained, a Leviathan (which in our age means an increasingly heavy-handed legal and administrative tyranny) is the paradoxical and bitter fruit of a polity based on his liberal but amoral principles.

Writing in the aftermath of the French Revolution, Edmund Burke articulated a political axiom that could have predicted this outcome:

> Men are qualified for civil liberty in exact proportion to their disposition to put moral chains upon their own appetites. . . . Society cannot exist unless a controlling power upon will and appetite be placed somewhere, and the less of it there is within, the more there must be without. It is ordained in the eternal constitution of things, that men of intemperate minds cannot be free. Their passions forge their fetters.[10]

In other words, a limited government compatible with wide personal liberty requires a virtuous people, a point well understood by the framers of the American Constitution. As John Adams said, "Our constitution was made only for a moral and religious people. It is wholly inadequate for the government of any other."[11] James Madison extended this understanding to all of politics: "To suppose that any form of government will secure liberty or happiness without any virtue in the people, is a chimerical idea."[12] In the end, living legally rather than morally is not desirable on political grounds alone: a lack of virtue in the people entails a government of force, not consent.

If we now turn our attention to humankind's relation with the natural world, the case for placing moral chains on human will and appetite becomes even more compelling. When Hobbes "unleashed the passions," he liberated men and women from imposed moral or religious strictures, but he also gave birth to what we know as economic development. Although the state no longer had the duty or even the right to inculcate or enforce private virtue, it did nevertheless have a positive role beyond mere peacekeeping—to foster "commodious living." Freed of the obligation to promote otherworldly ends, the state would henceforth dedicate itself to the things of this world—to abetting human desire, especially the urge for material gratification.

Following in Hobbes's footsteps, John Locke and Adam Smith made this profound shift in orientation from sacred to secular explicit: the purpose of politics is to facilitate the acquisition of private property and national wealth, along with the power that they confer. But the unfortunate side effect of unleashing human will and appetite in this fashion has been the destruction of nature.

Nature may not be a moral agent in the usual sense of the word—although a moral code is indeed implicit within the natural order—but it does have physical laws and limits that cannot be transgressed with impunity. Tragically, in the absence of mores that promote self-restraint and respect for nature, the exploitation of the natural world is bound to turn into overexploitation, for human wants are infinite. The long-term effect of unleashed passions therefore has been to violate nature's laws and limits and provoke an ecological crisis.

Our escalating ecological problems have become both common knowledge and a growing focus of political concern but to very little effect. After all, our form of politics requires perpetual economic growth, so the idea of limits, much less retrenchment, is anathema. Besotted with hubris, we cherish the delusion that we can overpower nature and engineer our way out of the crisis. We are not yet ready to admit that the destruction of nature is the consequence not of policy errors that can be remedied by smarter management, better technology, and stricter regulation but rather of a catastrophic moral failure that demands a radical shift in consciousness.

The antidote to political corruption and ecological degradation is therefore the same—a moral order that governs human will and appetite in the name of some higher end than continual material gratification. For this we need true laws, not merely prudent or expedient rules. But where shall we find such laws? They will not be found in revealed religions, old or new. Whatever the virtues and advantages of premodern religious politics, the concomitant evils and disadvantages were enormous, and Hobbes's philosophical revolt was both intellectually and historically justified. Perhaps they can be found in some new ideology? Again, surely not. If the history of the twentieth century has anything to teach, it is that secular ideologies are even worse than religious creeds at fomenting cruelty and violence. This leaves only one possible source for a new moral code—natural law, the law "written on the tablets of eternity."

The classic definition of natural law by Cicero is still unexcelled, albeit in need of minor amendment:

True law is right reason in agreement with Nature; it is of a universal application, unchanging and everlasting; it summons to duty by its commands, and averts from wrongdoing by its prohibitions. . . . We cannot be freed from its obligations by Senate or People, and we cannot look outside ourselves for an expounder or interpreter of it. And there will not be different laws at Rome and at Athens, or different laws now and in the future, but one eternal and unchangeable law will be valid for all nations, and for all times.[13]

Natural law in this sense has lost almost all philosophical respectability in modern times. As mentioned in the prologue, the Enlightenment *philosophes* believed that there was a discernible moral order in the cosmos that science would soon reveal. So they believed that natural law was not to be found by reflection alone but that we could—and indeed should—"look outside ourselves" to learn from nature. And this was a necessary correction, for the danger of a purely introspective quest for natural law is that we may mistake culture for nature. For example, the vast majority of Europeans and growing numbers of Americans condemn legal execution, but Confucians by and large do not. Abolition of the death penalty may therefore be a laudable moral goal, but it is probably not a candidate for natural law. Reference to some external standard—such as science, including the softer human sciences—can therefore provide a check on ethnocentrism and assure us that what we discover will be "valid for all nations, and for all times."

Unfortunately, the evolution of science took a very different path than the *philosophes* intended—away from a more inclusive reason and toward an increasingly narrow and instrumental rationality. Applying the methods of the so-called hard sciences to human affairs, the rationalists demonstrated (to their satisfaction) that there was no epistemological stance from which to derive natural law or moral principle. Instead of a sentient universe charged with moral meaning, science found only a machine—dead matter to be exploited by economists and engineers to make us wealthier and more powerful, not better. And what does a machine governed by mathematical formulas and physical laws have to teach us about how we should live? Nothing.

This brings us to the impasse at which we now find ourselves. As naked self-interest turns liberty into license and spreads demoralization, an increasingly despotic state tries vainly to forestall moral and ecological self-destruction with stopgap measures and ill-considered laws that cause mostly more harm than good.

A way out of this impasse has now emerged. As noted in the prologue, the epistemological and ontological revolution of the twentieth century has decisively overthrown the mechanical worldview and opened a path to the goal of the *philosophes*. By discovering and appreciating the moral order implicit within the natural world, we can derive ethical principles that will serve as a basis for polity and society in the twenty-first century and beyond. These principles are nothing new. The wise of every age and tradition have urged them on those who had ears to hear. The difference now is that what once was merely sensible has become imperative if a complex civilization is to survive.

But is it really possible to discover an ethical basis for politics that accords with natural law? The next three chapters argue that nature does indeed instruct us in how to live. Ecology, physics, and psychology—that is, biological nature, physical nature, and human nature—reveal fundamental and eternally valid moral principles with which to reconstitute our polity. On this virtuous foundation I shall try to construct a new Aristotelian rule of life whose essential core is a politics of consciousness dedicated to the idea that ennobling human beings matters more than accumulating dead matter.

The Sources of Natural Law

2
Ecology

All the errors and follies of magic, religion, and mystical traditions are outweighed by the one great wisdom they contain—the awareness of humanity's organic embeddedness in a complex natural system. And all the brilliant, sophisticated insights of Western rationalism are set at naught by the egregious delusion on which they rest—that of human autarchy.
—Phillip Slater[1]

The Stoic philosopher Zeno taught that the goal of life is to live "in agreement with nature."[2] But how can we know what this means? Those who hold tenaciously to a strict rationalism would argue that it is not possible to live in agreement with nature because there is no scientific way to get from *is* to *ought*, from facts to values. Similarly, almost all modern philosophers follow Nietzsche in maintaining that there are no facts, only interpretations. The languages of good and evil spoken by diverse human cultures therefore have no truth value; they are simply expressions of prejudice or the will to power. So before approaching ecology for moral instruction, we had better justify the quest itself lest what follows be dismissed out of hand.

According to Aristotle, different domains of discourse entail different degrees of exactitude.[3] Mathematical theorems demand ironclad proof, and scientific experiments require rigorous design and strong inference. But ethical questions have no precise answers—because discussions of human affairs can never be scientifically exact, only more or less reasonable. The most that we can hope for is a thorough and open-minded analysis of the best available information followed by a considered judgment as to the best course to follow. Yet this exercise of judgment is not contrary to the scientific spirit. As Albert Einstein pointed out, "Ethical axioms are found and tested not very differently from the axioms of science. Truth is what stands the test of experience."[4]

In short, assuming that morality is not simply given by revelation, it can be only reasonably discovered, not rationally proven. The hardcore positivist who insists that rationality—which is only one aspect of reason, however useful and necessary it is in its proper sphere—is reason itself and must always have the final say on every issue is therefore an epistemological fanatic, which is to say irrational.

The positivist is all the more irrational today in that the epistemological ground has shifted, rendering the extreme rationalism of the mechanical worldview obsolete. But every change in epistemology implies an equivalent change in ontology and also in ethos. The justification for the seemingly quixotic quest for natural law we are about to undertake will emerge from the quest itself. As is shown below, science no longer supports the rationalistic epistemology and ontology that has fostered an amoral, value-free ethic. On the contrary, the new postmechanical worldview logically entails a radically different ethos.

My presentation is necessarily selective and eclectic. It is not a textbook or a report from the research frontier but an effort to extract philosophical meaning and political guidance from ecology, physics, and psychology. To this end, I have tried to discover the metaphors that best convey the philosophical essence of each domain and that most clearly elucidate their moral import in layman's terms.

But this quest for natural law cannot lead us to absolute truth. As Aristotle says, perfect knowledge is not achievable concerning practical affairs or ethics. We must use our best judgment. My aim is not to build an airtight scientific case but to demonstrate a reasonable scientific basis for the political argument that follows. (In fact, there may be no such thing as an airtight scientific case. Despite the rigor of its methods, even scientific knowledge is never absolute. Science is a work in progress that occasionally undergoes revolutions and that sometimes discovers the extraordinary, such as the "dark" matter and energy that make up most of the universe.)

Epistemological rigor must yield to common sense and the practical needs of society. If an irrational rationality produces widespread social anomie and extensive ecological damage and also leaves us with nothing but technique for guiding human affairs, then a more judicious reason would seem to be essential, if only for prudence's sake.

As I argue later, the Platonic predicament is inescapable. Once we leave behind the realm of pure engineering, we are stuck with social fictions or lies. These are not true or false in the scientific sense, only more or less noble, beautiful, and useful. But these fictions cannot fly in the face of the facts, which is why I rest the political argument on a scientific foundation.

With this brief justification of my heuristic method, let us turn to ecology to see what it can teach us about how to live in happy agreement with nature. The liberation of man from nature is both the heroic virtue and tragic flaw of civilization. It makes possible the great material and cultural achievements that are synonymous with civilization, but it also foments the evils previously enumerated—and the greater the achievements, the greater the evils. The tragedy of modern industrial civilization resides in its greatness. All previous civilizations have exploited the natural world, usually self-destructively, but they never attempted to deny natural necessity or to stand apart from nature, much less over it. By contrast, industrial civilization has gloried in its ability to bend the so-called external world to its will.

This will to power over nature is the essence of modern hubris—an overweening end that is pursued by excessively rational means and driven by irrational urges. René Descartes and Francis Bacon, two of the principal authors of the modern way of life, regarded nature as a hostile power to be dominated without ruth or scruple.[5] Sigmund Freud, the last great defender of the Enlightenment (despite his own rediscovery of the irrational), put his finger on the neurotic origin and character of this hostility: "Against the dreaded external world one can only defend oneself by . . . going over to the attack against nature and subjecting her to the human will."[6] So modern hubris originates in irrational dread and manifests as an unlimited war against nature for wealth, power, and dominion.

Preserving the environment is thus the lesser part of the problem. Industrial civilization must indeed stop abusing nature and depleting resources before it follows previous civilizations in committing ecological suicide.⁷ But the only real solution is to put an end to the hubris itself by dissolving the dread-driven, neurotic hostility to nature that fuels the urge for domination.

Ecology is the surest cure for modern hubris. To understand ecology is to see that the goal of domination is impossible—in fact, mad—and that the crude means we have employed to this end are destroying us. To understand ecology is also to see that some of the most vaunted achievements of modern life—our extraordinary agricultural productivity, the dazzling wonders of technological medicine, and, indeed, even the affluence of the developed economies—are not at all what they seem but instead are castles built on ecological sand that cannot be sustained over the long term. In short, ecology exposes the grand illusion of modern civilization: our apparent abundance is really scarcity in disguise, and our supposed mastery of nature is ultimately a lie.⁸

To put it more positively, ecology contains an intrinsic wisdom and an implied ethic that, by transforming man from an enemy into a partner of nature, will make it possible to preserve the best of civilization's achievements for many generations to come and also to attain a higher quality of civilized life. Both the wisdom and the ethic follow directly from the ecological facts of life: natural limits, balance, and interrelationship necessarily entail human humility, moderation, and connection.

Like any other species, *homo sapiens* is subject to natural limits. Technology does give human beings an ability to manipulate the environment that other species mostly lack. But humanity's success in this regard is in large part illusory because it has been purchased at a high price—symbolized by the accelerated extinction of those other species, with all that this implies for our own long-term future.

Technological man has neither abolished natural scarcity nor transcended natural limits. He has merely arranged matters so that the effects of his exploitation of nature are felt by others. Other species, other places, other people, other generations suffer the consequences of the intensified ecological imperialism of the modern age. The current environmental problematique testifies to the impending failure of this strategy.

The limits on human action are physical, biological, and geological but also systemic. Reserving a fuller discussion of complex adaptive systems governed by a multiplicity of interacting feedback loops for the next chapter, I simply note here that the biosphere and all its subsidiary ecosystems are characterized by nonlinear dynamics that make them difficult to understand and harder to control. In fact, we cannot really know what the ultimate limits are.

To put it the other way around, just as games are constituted by the rules that regulate play, the limits themselves constitute natural systems. To be without limits is to be without structure and therefore to be entropic—chaotic, useless, or unintelligible. And limits do not oppose freedom: "Structure and freedom," says Jeremy Campbell, "are not warring opposites but complementary forces."[9]

To attack limits is therefore foolhardy, for it risks destroying the system. And to aim at control is ill-advised, if not impossible. From the systems point of view, the soundest and safest strategy is not control but cooperation—accepting limits and working within them to achieve reasonable human objectives rather than seeking domination and riches at the expense of the system.

Limits, however, are an affront to the self-image of modern man, who believes he is master of all he surveys and can act as he likes without consulting the rest of creation. To accept that the human species is but one small part of an organic web of life that places fundamental constraints on our actions is bitter medicine indeed for the heirs of Bacon and Descartes. But it seems that we shall have to swallow the medicine nonetheless, abandoning the delusion of radical separation that fuels the illusion of unlimited mastery.

Humility is therefore the essence of ecological wisdom and the foundation of an ecological ethic. Not only are limited natural systems opposed to unlimited human appetites, but limits also oblige us to come to moral terms with the web of life—that is, to renounce hubris and to find a place within nature instead of above it. From this follows a duty to deal justly with the nonhuman world as well as our own posterity.

Once we abandon our anthropocentric point of view, we can see limitation as a creative force that fosters quality instead of quantity. The life process responds to limited matter and energy by using them more efficiently. The effect is to create richer and more complex ecosystems:

Picture the difference between a field of weeds and a tropical rain forest: the former has a few species of plants and animals that grow rapidly and die off equally rapidly, leaving little behind to mark their passage; whereas the latter exhibits a rich organic tapestry of many diverse life forms mutually interacting with each other in complex ways. Depending on how it is organized, therefore, the same limited budget of solar energy can produce strikingly different outcomes: a brief and flashy display or a rich and enduring stock.[10]

This contrast between so-called pioneer and climax ecosystems—between the weeds and the rainforest—indicates the direction in which civilization needs to evolve. Modern civilization is hell-bent on weedlike quantitative growth, and that must change. When we have become humble enough to work with natural limits instead of against them, we shall use matter and energy more frugally and create a qualitatively better climax civilization.[11]

The corollary of natural limits is natural balance. The organic world is a complex, interconnected living system that has its own autonomy, integrity, and value. To treat it as though it were simple, divisible, and dead—that is, as though it were a machine to be manipulated at will—shows a fatal lack of understanding.

Natural balance is not a simple or fixed state, for living systems are fundamentally dynamic. Change and adaptation are almost the only constants. Yet out of life's fertile dynamism emerges a surprising stability, maintained by feedback mechanisms that keep living systems in homeostatic balance as they fend off disturbance and dance between order and chaos.

Unfortunately, human action works against natural homeostasis. In their natural state, for example, ecosystems respond slowly and minimize throughput, features that promote stability. By contrast, economic man favors "profitable" systems—such as fields of fast-growing hybrid corn or herds of hormone-laden cows—that have fast responses and high flow rates. They are high-yielding but intrinsically unstable and therefore incapable of surviving without constant intervention. In other words, one cannot have *both* high yields *and* homeostatic stability; the former precludes the latter.

In addition, although natural systems are resilient when subjected to one-time insults, chronic stress causes progressive damage because the constant pressure of even low-grade stress inhibits or frustrates the forces that promote homeostatic recovery. Just as the human body can tolerate the occasional overindulgence without serious harm but succumbs to the accumulated effect of bad habits, so too the steady drip of man-made chemicals into the environment causes ecosystems to sicken and die.

In practical terms, this means that no part of the system can be allowed to maximize its own gains, for this would destroy the balance of the whole. By their nature, living systems always optimize: they seek the happy medium that best accommodates the interests of the whole and the parts. It follows that humanity's attempt to maximize its wealth and power at the expense of the rest of creation is fundamentally antibiological and self-destructive. Ecology suggests—nay, ordains—optimization as the proper strategy for long term survival and well-being.

This is tantamount to saying that ecology validates the golden mean. And from the wisdom of "nothing in excess" follows an ethic of "moderation in all things" as the only way to preserve and respect the autonomy, integrity, and value of the whole of which humanity is simply a part.

To speak of limits and balance is to acknowledge the fact of natural interdependence, which leads to the deepest ecological wisdom and the highest ecological ethic. The life process is a unity in which everything is connected to everything else. Life is simply one very large ecosystem—the biosphere—made up of progressively smaller ecosystems that nest inside each other hierarchically to form a chain of being from higher to lower. At the same time, these systems also interrelate with each other nonhierarchically to form a complex and manifold web of life in which humanity is organically embedded.

Our view of the life process is distorted. We tend to focus on our own species and a few of the larger animals, whereas the bulk of the earth's biomass consists of tiny creatures, most of which are invisible to the naked eye. Asked what a lifetime of biological studies had taught him, J. B. S. Haldane is said to have replied that the deity had an inordinate fondness for beetles. But the great number and variety of beetles and other insects pales beside the multiplicity of microorganisms that are the true creators and sustainers of life on the planet. Without them at the base of the pyramid, none of the so-called higher forms of life would exist.

Indeed, there is no such thing as an individual life because organisms cannot by themselves sustain life. For their sustenance, they depend completely on the whole system—a system made up of many organisms of different species interacting with their physical environment to produce the flow of chemicals and energy that individuals require to stay alive. Without this support by a community of the living and nonliving, the individual organism simply has no existence.

In fact, every so-called individual is really a symbiosis, a collectivity of cells. And cells are themselves symbioses: they are composed of organelles (specialized cell parts that resemble and function as organs) that are the direct descendants of previously free-living bacteria and algae that joined forces eons ago to create the little biospheres that are the common building blocks of all higher life forms. In other words, without the organized cooperation of billions of tiny creatures living together symbiotically, human beings would not exist. It took eons of biological cooperation to create complex creatures like us; it takes a vast symbiotic effort to sustain us now that we are here.

So although competition and predation are indeed brute facts of natural life, mutuality and cooperation predominate. From our exalted perch, we have a foreshortened view of nature as red in tooth and claw, but this is not its essence. The overall strategy of evolution is to incorporate, not exterminate. Evolution is really coevolution, for nothing has evolved in isolation. Instead, each species has been selectively shaped to fill a certain niche in the web and to be a fitting partner to other species, as exemplified by the bees and the flowers. Even the competition between predator and prey involves a kind of cooperation, for each serves to perfect the other (and thereby produce such marvels as the Serengeti).

As a consequence, both within ecosystems and within the biosphere as a whole, evolution tends toward the climax—that is, toward a luxuriance of mutualistic symbioses. The result is more of life—vastly greater richness, complexity, order, and beauty than could ever be achieved if the process were merely competitive. If nature could be said to have an ethos, it is mutualism—harmonious cooperation for the greater good of the whole that simultaneously serves the needs of the parts.

What is astonishing is that this ordered profusion of life created itself: it self-organized. The Gaia hypothesis put forward by Lynn Margulis and James Lovelock states that the biosphere is, in effect, a kind of giant cell created by life forms that have adapted the chemistry and geology of the planet to their own purposes. So earth is not a dead rock with a tiny film of life on its surface but a living planet that incorporates even the rocks into the life process. Although it is technically not an organism (because it neither eats nor excretes), it is nevertheless "the largest organic being in the solar system."[12]

In short, earth has a physiology and a metabolism—it "lives" by "eating" solar energy and "excreting" waste heat—and is therefore, to that degree, alive.

To spell this out in a little more detail, through a long process of coevolution, organisms have interacted both with themselves and with their physical environment to transform the planet. Fueled by solar energy, they fostered the emergence and then the further development of an oxygen-based life process by creating the containing membrane of the atmosphere and the sustaining medium of the soil. As a result, air, sea, and earth now shelter and nourish a vast community of beings who are all genetically related and who are all linked together in obligate ecological relationships.

Having created such unlikely conditions—conditions far removed from the expectations of chemistry and also from those originally prevailing on the planet—living beings now continue to maintain them through complex chemical, biological, physical, and even geophysical feedback mechanisms. The earth is kept in a state of homeostasis. The parameters critically important to life—the gas balance, the level of salt in the oceans, global temperature, and so on—are kept at near constant levels by unremitting organic activity. Without this incessant work to maintain homeostatic stability against the tug of entropy (the universal tendency toward disorder), critical imbalances would soon occur, and virtually all forms of life would perish. Chopping down entire rainforests or dumping large quantities of chemicals into the oceans and atmosphere is asking for trouble. Such actions menace the physiology of the planet, a topic to which I return in the next chapter.

Although the Gaia hypothesis provoked considerable controversy at first and the word *alive* still sticks in the craw of many scientists, the facts are now relatively settled science, albeit purged of the taint of teleology that occasioned the original controversy.[13] It is not necessary to invoke notions of purpose or design to explain the existence of Gaia, any more than we have to believe that human beings are the product of divine intervention instead of natural evolution. Gaia is an "emergent property" of the earth's self-organizing systems. When Lewis Thomas asks himself what the earth most resembles, he answers, "It is *most* like a cell," and is therefore a living being: "Viewed from the distance of the moon, the astonishing thing about the earth, catching the breath, is that it is alive."[14] Similarly, the ecologist Daniel Botkin, while denying life to the earth with one breath, gives it back with the other: "The Earth is not alive, but the biosphere is a life-supporting and life-containing system with organic qualities, more like a moose than a . . . mill."[15]

What is even more astonishing and significant is that mind coevolved with matter—or is it the other way around?—because cognition is inherent in self-organization. Self-organizing systems, like Gaia or the human body, may not think, plan, or decide like human beings, but they still "know" how to keep themselves balanced and healthy. In other words, although most such systems are not capable of intellectual activity, they nevertheless perceive, respond, choose, remember, learn, adapt, and even invent. Thus, said biologist Humberto Maturana, "Living systems are cognitive systems, and living, as a process, is a process of cognition. This statement is valid for all organisms, with and without a nervous system."[16] Parmenides put it more pithily: "Thinking and being are one and the same."[17] So consciousness is intrinsic to life. Comparatively weak in the most primitive organisms, it gradually grows stronger and more intelligent as nervous systems become more complex, until it evolves into symbolic and self-reflexive thought.

We humans tend to overvalue the latter and to look down on the rest of the continuum, but to think that only we are truly conscious is to compound ignorance with arrogance. The vaunted reasoning mind stands, as it were, on the shoulders of all the lesser forms of consciousness that exist within the little human biosphere—the mammalian, the reptilian, the cellular, and even the molecular and atomic. In addition, the distinction between humans and the allegedly lower animals is not that we possess unique capacities and superior abilities. This is true of every species, no matter how humble. Ants, for instance, taste their way through life with a speed and accuracy that would shame a whole laboratory of organic chemists. Nor is the distinction simply a question of some species being hypersensitive. We are everywhere confronted with entire modalities of sensing and communicating well beyond what is normal for humans, such as the rich sonic world of the cetaceans. And animals are not mere instinct-driven automata: bumblebees are "free agents" in deciding where to forage, so they do indeed appear to have a "mind" to make up.[18] Above all, mere microbes were quite intelligent enough to create the basic geochemical cycles and processes on which all the so-called higher forms of life depend.

More to the point, animals higher on the evolutionary tree are closer to us both genetically and even behaviorally than we once believed. The more attentively and sympathetically we study mammals and even birds, the more we discover that their perceptual, cognitive, and emotional lives are far richer and more complex than we used to believe and also that they exhibit a moral nature.[19]

This accumulating body of zoological knowledge presents an epistemological challenge to humanity's concept of reality and to any pretension to superior knowledge of it based on scientific investigation. If each species perceives only a part of the full spectrum of reality and then interprets that part in its own characteristic way, whose reality is ultimately real? The honeybee's world of color and calculus? Or the human's world of sound and geometry? In fact, each species selects from the total spectrum of available sense data the little information necessary to construct the particular reality essential to its survival.

Because each species is intelligent in its own fashion, the kind of mind called *nous* by the Greeks—that is, an intelligent ordering principle or force—is found in all forms of life, even the most primitive; and it exists in both the organisms themselves and also the ecosystems that enfold them. In this light, it does not seem totally outrageous to hypothesize that the biosphere, a kind of cell writ large, is alive and conscious in some fashion. So we have come full circle: like the ancients, we may once again come to regard the earth as a living creature endowed with soul and reason.[20]

Nature is not a machine. Nor does humanity stand apart from nature entitled by evolutionary right to lord it over the earth. In consequence, the modern way of life and politics based on the mechanical worldview is rendered obsolete, both philosophically and practically.

In politics, for example, a fundamental principle of classic liberal political philosophy is that individual freedom ends where harm to others begins. In practice, this has not greatly inhibited us because another liberal tenet says that individuals are intrinsically separate from each other and from nature. Thus only egregiously antisocial or antiecological behavior qualifies as harm. But if pervasive interdependence makes us an inseparable part of the common stream of life, then the liberal fiction of separation collapses. There are no truly private decisions. Whatever I do to the stream affects all of life, including myself. Behavior hitherto regarded as legitimately selfish becomes overtly harmful—and therefore morally reprehensible, even by liberal standards.

In effect, ecology teaches an ancient wisdom—the wisdom of the great chain of being, albeit in a new and different form. Nature is more web than chain, and the wisdom is grounded in science rather than theology or revelation. Eastern metaphors, such as the Tao or Indra's Net, may therefore be better representations of ecological reality. But the ethic that follows would gladden the heart of a saint: love creation, and regard all beings as brothers and sisters.

So humanity's organic embeddedness changes everything. In effect, the political challenge before the human race is to secure both its dignified survival and its further moral development by including all of life in the process of governance. It is to ensure that the interests of all creatures and all generations are taken into account by the political process. Once we acknowledge our deep connection to the whole of life, then compassion, equity, freedom, justice, prudence, and even reason appear in a radically new light.

Take our current notion of justice as one example. For us, justice is mostly about due process—that is, a juridical mechanism guaranteeing fair play among those now living. But the interests of justice in the larger sense of moral rightness or equity are not always well served by such a narrow definition. Among other things, the interests of posterity have no legal standing and are therefore almost entirely disregarded. An ecological concept of justice would be very different, for it would extend in time as well as space. If we take into account the evolutionary development of the life process, then we are obliged to acknowledge with Burke that the present generation has received the past as a patrimony that it holds in entail for future generations. We who are now living have been entrusted with the ecological capital of all the ages. To fail to appreciate its beauty and respect its provenance is barbarism. To waste and consume it without regard for posterity is vandalism.

To mention Burke is to see that ecology, because it is grounded in evolution, has fundamentally conservative political implications. A long process of trial and error has weeded out the bad innovations, leaving behind what has stood the test of time. The result may not be perfect, but it is probably the best that can be accomplished with the materials at hand. Evolution or ecology should not be used to justify wealth and privilege or inherited evils, but it does imply a Burkean stance toward change. There is a kind of wisdom contained in the system—biological or social—that we would be wise to study and understand before we launch "reforms" based on our "progressive" ideas. Among other things, as the previous chapter illustrates, replacing common law that is developed over many centuries by disinterested jurists with positive or fiat law that is made up out of whole cloth by partisan legislators is unwise to the point of folly.

An additional problem with defining justice in this fashion is that it both reflects and reinforces the strongly individualistic character of modern society. A legal system based on due process endows individuals and corporations, separate parts of the whole, with rights that too often contravene the common interest—as when private entities profit at the expense of the commons by depleting or polluting or when parents refuse to immunize their children despite the burden this imposes on public health.

From the ecological perspective, elevating the individual over the community makes no sense, intellectually or practically. Individuals should have civil rights, but civic duties and responsibilities must have at least equal weight if we wish to preserve the integrity of the system over the long term.

Indeed, a climax ecosystem closely resembles Plato's ideal state. A truly ecological notion of justice would be revolutionary. If the biosphere is a vast web of interconnecting species, with each occupying a particular niche within the larger whole so that each thereby contributes to upholding the beauty, complexity, and richness of the whole, then might not the same be true of the human sphere? Perhaps human survival and well-being would be better served by a more Platonic notion of justice—in other words, a polity in which individuals fulfill social roles rather than aggrandize private interests. We return to this and other controversial issues later after we have seen how physics and psychology are also haunted by the shade of Socrates.

To conclude, because the evils peculiar to modern civilization derive largely from its fundamentally antiecological principles, ecology is the necessary cure for them. To be blunt, modern political economy is contradicted root and branch by ecology. Even well intentioned efforts to "green" that economy only deepen the contradiction because they proceed from the same mindset that created it in the first place.[21] Future historians will probably regard our era as an age of collective folly—a time when large numbers of highly intelligent and influential people fervently believed (despite Adam Smith's clearly expressed sentiments to the contrary) that society and polity should also be ruled by the same invisible hand that governs a market economy. When we awaken to this folly, the wisdom and ethic of ecology will become the intellectual and moral basis of a reinvented political economy capable of preserving civilization over the long term. Among other things, the reconciliation of man and nature implicit in an ecological way of thought and life would restore the coherence and meaning lost when man made nature into an enemy. Orphaned no longer, he would once again be at home in the universe. For all these reasons, ecology will have to be the master science and guiding metaphor of any future civilization.

3
Physics

> Today there is a wide measure of agreement . . . that the stream of knowledge is heading towards a non-mechanical reality; the universe begins to look more like a great thought than like a great machine. Mind no longer appears as an accidental intruder into the realm of matter; we are beginning to suspect that we ought rather to hail it as the creator and governor of the realm of matter.
> —Sir James Jeans[1]

Every age has a master science that constitutes what might be called its paradigm of paradigms—an overarching metaphor for the pattern that all other theories, scientific and nonscientific, will follow. The character of that science has profound political, social, and economic consequences. The master science of antiquity was astronomy. It elevated the sky gods to the throne of heaven and raised the pharaohs, the semidivine surrogates of the gods, to the apex of the political pyramid. It also modeled the seasonal regularity on which agrarian civilizations depended. Similarly, when the scientific revolution inspired by Descartes and Newton made exact-law physics into the new master science, it also created the modern age as we know it—an age of mechanism, reductionism, determinism, materialism, individualism, and so forth, all obedient to the essential character of that science.

For the followers of Descartes and Newton, the world was a kind of automaton. Implicit in this view was the idea that there is an objective reality "out there" that is completely accessible to observation and reason. It should be possible, in principle, to discover the whole truth about that external reality and thereby achieve a God-like position above nature allowing us to become its master and possessor.

But physics is no longer what it was in Newton's day. Revolutionary advances during the twentieth century have transformed our understanding of the universe and the process by which we know it. Consider the epithets used to characterize modern physics—relativity, nonlocality, uncertainty, indeterminacy, and so on. Or ponder the awkward paradoxes it has generated—that general relativity and quantum mechanics are both experimentally true and theoretically contradictory, that light is both wave and particle (but not exactly either), and that Schrödinger's cat is at the same time both dead and alive. Even mathematics is now known to be "incomplete" (because it is bound to take something for granted). Nor is geometry invariable. Euclid's way of describing space is only one among many (and by no means the most important). In other words, we no longer inhabit a fixed, certain, stable, ultimately knowable clockwork universe. Far from being an automaton, the cosmos appears more and more like a work of art, just as it was for the ancient Greeks.

It is not that nature is not lawful or that anything goes. But when there is no longer an objective reality that is fully accessible to observation and reason, our previous pretensions to all-knowingness and consequent mastery over nature are set at naught. In addition, because we are embedded in that nonobjective reality, mind now intrudes into matter. In the oft-quoted words of James Jeans, the universe appears "more like a great thought than a great machine." It seems that nature is fundamentally Platonic: what we experience, both with our perceptual apparatus and with our scientific instruments, is but a shadow cast on the wall by a deeper level of reality—a reality that transcends space and time as commonly understood.

Paradoxically, we carry on as though none of this had happened. Politics, law, economics, academe, and daily life are dominated by obsolete ideas—for example, simple-minded empiricism or fixed concepts of space and time—that science was forced to abandon almost a century ago. Even most practicing scientists cling to a notion of the scientific enterprise that is inconsistent with the new understanding. Despite having lost its original theoretical basis, an outmoded mechanical worldview still prevails.

The resulting intellectual muddle has serious consequences—not just in the realm of theory but also in practice—because what is defined as real by human thought is inevitably made real by human action. A natural world conceived of as dead and treated as such eventually becomes dead in fact. Similarly, when nature is taken to be essentially atomic, the result is political and economic doctrines that grind men and women into social atoms.[2] And if material reality is all that exists, then spirit and instinct are meaningless categories slated for the dustbin of history. The totality of modern life is therefore the concrete expression of a particular notion of reality—of a metaphor that is both dead and deadening—and all of its contradictions and problems trace back to the falsity of that notion.

The new view of nature is radically different. Physical reality is a reflection—a play on our senses—of one all-encompassing and interconnected cosmic process. Everything from the tiniest quark to the largest galaxy, whether it is what we call living or nonliving, is simply one aspect of an undivided whole.

We can distinguish conceptually among the various aspects—the chromosome from the cell, the bee from the flower, the solar system from the galaxy—but they are not truly separate. They are only particular projections in space and time of the underlying process—mere phenomena or, quite literally, "appearances" that emerge from nothingness into existence and then dissolve back into nothingness in obedience to the laws of nature that govern microcosm and macrocosm alike.

The cosmic process is often compared to a wild dance of creation and destruction. As the universal flux vibrates to the harmony of nature's laws, fleeting patterns arise in space-time, flourish briefly as seeming objects, and then vanish into the void whence they came. Even during their evanescent existence, the objects are not solid entities but instead a blur of whirling particles. This dance is invisible to our senses, which make us experience a stable word composed of solid material objects. However, now that we understand the mathematical music to which the dance is performed, we know that our common-sense view is illusory: there is only the dance.

So reality is not made up of objects, nor is it composed of space, time, matter, energy, light, or anything else. There is no basic "stuff" of the universe, and phenomena are inherently ephemeral, mere standing waves in the stream of existence.

Physical reality is essentially insubstantial. It is composed of sets of fields governed by physical laws, and everything else is derivative, a mere artifact of the dynamics of these fields.[3]

But if what we perceive with our senses and instruments is ephemeral, insubstantial, and derivative, then we are not really in touch with an objective reality. Einstein captured our epistemological plight with a delightful metaphor: "Nature shows us only the tail of the lion."[4] Unable to see the lion directly, physicists twist its tail this way and that, inducing experimental roars that are analyzed to produce equations—that is, a mathematical description of the beast. So they know the lion by inference, not observation.

Our epistemological plight has worsened in recent years with the discovery that "ordinary" matter, which we once took to be all of matter, is only about 5 percent of the universe by mass. The rest is invisible "dark" matter and energy, about which we know next to nothing except that they have to be there (because we can observe their gravitational effects). Were Einstein living today, he might have to say, "Nature shows us only the tail of the lion and hides from us completely the vastness in which he dwells." But if reality is not actually found, only inferred, then the physicists' picture of it is subjective. The picture is not arbitrary or fantastic, but it necessarily reflects what is in the mind of physicists to begin with—the theories and methods that they use to investigate nature. "We have to remember," said Werner Heisenberg, "that what we observe is not nature in itself but nature exposed to our method of questioning."[5] Erwin Schrödinger spelled out the profound implication: "Mind has erected the objective world of the natural philosopher out of its own stuff."[6] Arthur

Eddington made the subjectivity explicit by saying that "the solid appearance of things . . . is a fancy projected by the mind into the external world."[7]

Heisenberg drew the necessary philosophical conclusion: "Modern physics has definitely decided for Plato." Why? Because "the smallest units of matter are . . . not physical objects in the usual sense of the word; they are forms, structures or—in Plato's sense—Ideas, which can be unambiguously spoken of only in the language of mathematics."[8] Eddington concurred: "The exploration of the external world by the methods of physical science leads not to a concrete reality but to a shadow world of symbols, beneath which these methods are unadapted for penetrating."[9] Likewise Jeans: "Imprisoned in our cave, with our backs to the light, [we] can only watch the shadows on the wall," so all of our descriptions of nature are mere "parables."[10]

Physics therefore joins ecology in seeing *nous* in what we once believed was dead matter. Echoing Schrödinger, Eddington called matter "mind stuff."[11] "Matter," says Freeman Dyson, "is not an inert substance but an active agent, constantly making choices between alternative possibilities. . . . [Mind] is to some extent inherent in every electron."[12] Even the late Richard Feynman, famous for warning against leaping from quantum results to philosophical conclusions, nevertheless spoke almost mystically of "atoms with consciousness" and "matter with curiosity."[13] So we have again come full circle to an ancient understanding: "Astonishing!," exclaimed Pythagoras, "Everything is intelligent!"[14]

Nor is physics alone in deciding for Plato. Findings in many different fields—such as anthropology, linguistics, cognitive psychology, and depth psychology—tend toward a similar conclusion. The radical implications of particle physics cannot be dismissed as irrelevant to "real" life.

Studies of human perception, for example, show that the human brain and mind are not the simple and straightforward instruments for experiencing reality that the naïve assume. On the contrary, perception is actually cognition. From a vast sea of phenomena, we actively and selectively filter, like some mollusk, the information we require for day-to-day survival.

To begin with a few physiological facts, we do not "see" or "hear" anything. Our organs of perception are impinged upon by various pressures, vibrations, and other sensations. These are converted into electrochemical impulses, which travel along nerves to the brain, where they cause neurons to fire. The patterns of the firing, which are nonrepresentational, are then mathematically analyzed by the brain, using genetically determined programs, and converted into "images" or "sounds." In other words, the nervous system does not simply *register* impacts and intensities. It performs an *analysis* of the signals received and then constructs a *synthesis* according to principles inherent in the mind. This synthesis is what we actually "perceive."

Human perception is therefore the end product of a complex process that we can liken to a Hollywood studio operating at lightning speed to create a nearly flawless and utterly convincing representation of reality. Indeed, it is so flawless and convincing that we are not aware that everything in it has been selected, edited, and transformed by our physiology and our brain's method of operation and by bias due to personality, language, and culture.

To speak only of language, human beings inhabit a largely semantic reality made up of the linguistic categories of their culture. It is a closed loop. We tend to perceive only what we have words for, so what we have words for is what we habitually perceive. Our fundamental cultural assumptions are so deeply embedded in language that we are effectively imprisoned by them. As Ludwig Wittgenstein said, language frames our reality so completely that "the limits of my language mean the limits of my world."[15]

In addition, we tend to overlook the degree to which the possessive, aggressive, and territorial drives that have their seat in the "lower" brain influence higher-level mental processes. As the Indian proverb has it, "When a pickpocket meets a saint, all he sees are his pockets."

What we "see" is therefore a function of the nervous system's evolutionary history and of the long and tenuous chain of assumption and inference on which human perception rests. In effect, said neuroscientist Francis Crick, everything we perceive is in a certain sense "a trick played on [us] by [our] brains"—in other words, a shadow play, just as Plato maintained.[16] Yet the mind makes it seem as if we are in direct contact with reality, and we believe the deception. But the truth is otherwise: behind the visible forms lie invisible deep structures that engender them.

As noted previously, human perception is not the only product of invisible deep structures. Unseen fields generate all of physical reality—that is, the natural systems that constitute our world. And these systems have two characteristics that we must understand if we are to deal with them intelligently. Unlike the artificial, mechanical systems constructed by man, which are merely complicated rather than truly complex, natural systems are both self-organizing (or adaptive) and genuinely complex (or chaotic).

A self-organizing system determines its own structure and functioning to achieve long-term, dynamic stability within its environment. Compare a jet engine to a tiger. Although both are organized—and both are open systems that function by "burning" external matter and energy—only the tiger is *self-organizing*. That is, a tiger is capable of sustaining itself by itself in a challenging environment for as long as it lives. In addition, since it can make more tigers, there is a sense in which it never really dies. Individual tigers may pass away, but the race lives on (unless extirpated by man). The jet engine, by contrast, has to have its fuel supplied by an outside agent and is incapable of maintaining, much less reproducing, itself. After a few years of useful life, it goes onto the scrap heap, leaving no progeny.

Self-organizing systems are therefore relatively autonomous, stable, and enduring, whereas artificial systems are intrinsically dependant, unstable, and transient. Moreover, unlike machines, whose nuts and bolts can be only rearranged, self-organizing systems can adapt to changing circumstances and can even evolve into something more complex. Indeed, the theory of self-organization casts Darwin in an entirely new light. Evolution occurs not by natural selection alone but mostly as an outcome of the dynamics of self-organization. "Matter's incessant attempts to organize itself into ever more complex structures," says Mitchell Waldrop, produces an emergent order that is then honed by natural selection to produce the rich panoply of life, so a "deep, inner creativity . . . is woven into the very fabric of nature."[17]

Biological systems are not the only self-organizing systems. The chemist Ilya Prigogine received a Nobel Prize in 1977 for discovering self-organization in inorganic systems. If fluctuations disrupt a molecular system's stability, it may simply break down, but in some cases, it responds to the challenge by reorganizing itself to be more complex, robust, and stable. A similar process formed the stars and galaxies.

This deep, inner creativity leads nature ever onward toward greater sophistication, complexity, and functionality—or, in a word, quality—of evolutionary "design." In a way, self-organization is tantamount to a scientific version of vitalism. Life transcends mere matter, says Waldrop, "not because living systems are animated by some vital essence operating outside the laws of physics and chemistry" but because animation consistently emerges out of the dynamics of inanimate systems.[18] Again, *nous* is pervasive. Far from being dead, matter exhibits a kind of intelligence that drives it toward transcendence.

Unfortunately, there is a radical mismatch between complex, self-organizing natural systems and complicated, man-made mechanical systems. The latter require relatively large inputs of high-grade matter and energy for their functioning, and they also generate much larger outputs of useless or even toxic materials than self-organizing systems, which are typically quite thrifty (a point to be pursued in greater depth below).

What is worse, the relentless demands of humanity's artificial systems come too thick and fast for natural systems, which evolved to handle a different kind of challenge over a much longer time scale. The tiger is no match for a bulldozer that in a few days can knock down the millennia-old forest in which it dwells. And the atmosphere is probably no match for the engines that spew out in decades vast quantities of greenhouse gases that cannot be absorbed in less than ten thousand years.

This mismatch is a fundamental problem for humanity going forward. Self-organizing systems are intrinsically stable and resilient—but only up to a point. Chronic, low-grade stress sickens them, and once stressed beyond a certain threshold (sometimes called a tipping point), they can fall out of equilibrium and enter a regime of positive feedback in which a self-destructive process feeds on itself. For example, as warming global temperatures melt the permafrost, it releases long-frozen methane into the atmosphere. This potent greenhouse gas is then likely to accelerate the warming, causing more melting and releasing more methane, and so on, in a classic vicious circle. So many natural thresholds have been exceeded in recent years that positive feedback is driving many natural systems toward destruction. In consequence, the total world system is now in what theorists call an overshoot-and-collapse mode.[19]

We are just beginning to grasp the vital role of self-organization in creating biophysical reality and to develop the intellectual tools required to study it appropriately. But what we already know suggests that we should approach self-organizing systems with respect. As pioneered by Gregory Bateson and Erich Jantsch, the theory of self-organization is a synthesis of physics, evolution, ecology, systems, and cognition that terminates in an explanation of how mind emerges from—or, more properly, coevolves with—matter. Jantsch concludes by saying,

> Natural history, including the history of man, may now be understood as the history of the organization of matter and energy. But it may also be viewed as the history of the organization of information into complexity or knowledge. Above all, however, it may be understood as the evolution of consciousness, or in other words, of autonomy and emancipation—and as the evolution of the mind.[20]

In effect, self-organizing systems are more like autonomous beings than dependent machines and deserve to be treated accordingly.

The new science of complexity—or so-called chaos theory—penetrates even more deeply beneath the appearance of things to discover an astonishing regularity in phenomena hitherto believed to be random and unconnected. Wherever researchers look—for example, at turbulence in tiny cells, in swirling smoke, or in spiraling galaxies—they discover hidden principles of order within the apparently chaotic. Moreover, as the above examples suggest, these principles cut through all levels of reality. "The laws of complexity," says James Gleick, "hold universally, caring not at all for the details of a system's constituent atoms."[21]

Thus the ghost of Einstein is pacified at last: God does indeed play dice with the universe, but the dice are "loaded."[22] As we have already seen, the overall evolutionary tendency of matter and energy is toward greater complexity, sophistication, and intelligence. Nature definitely favors certain outcomes over others.

Chaos theory has partially resurrected the discredited idea of *telos*. There are definite ends toward which natural processes are "attracted," but contingency decides specific outcomes. Peatlands all over the world converge on a similar physical form, no matter where located or however evolved.[23] Yet, as Stephen Jay Gould has pointed out, if we were to run the tape of evolution again, the effect of contingency would alter the outcome, perhaps quite dramatically, even if major patterns remained the same.[24]

The fundamental nature of the order discovered by chaos theorists is not mechanical. Nature is nonlinear "in its soul," says Gleick: "the solvable, orderly, linear systems" taught to science students as the norm are, in fact, "aberrations."[25]

Nonlinearity is best understood by example. In a mechanical system, linearity prevails—that is, the link between cause and effect is clear and distinct. When coal is shoveled into a boiler to make steam to turn the wheels that propel a locomotive down the tracks, every link in the chain of causality can be traced and calculated. By contrast, when fertilizer is shoveled onto a field, its effect on eventual crop yield will probably be favorable. However, the outcome will depend on the amount applied, its chemical makeup, and a host of other factors—weather, soil, water, cultivation techniques, and so on—all of which will either enhance or counteract the effect of the fertilizer. Because each element of the system is related to all the others by multiple feedback loops, everything is simultaneously cause and effect, making it impossible to trace or calculate all the linkages. It is as if complex adaptive systems had a mind of their own: their behavior is autonomous and not fully predictable.

Such systems present formidable challenges to human understanding—and even greater ones to human intervention—because they are not solvable. They may be entirely lawful, but contingency and complexity conspire to frustrate determinism. Even if all important components, critical interrelationships, and governing laws are known with reasonable certainty—a major assumption—it may still not be possible to predict the behavior of even relatively simple nonlinear systems. How much more so if the system is far from simple—such as the global climate regime, which is an ecology of many interacting nonlinear systems.

Unpredictable does not mean unknowable, however. Although they cannot be solved, complex systems can be simulated to reveal their modus operandi. Unfortunately, although the laws that create complexity may be simple and few, the existence of many, many loops of mutual causation, or feedback, makes simulation heavily dependent on judgment about what to include or assume in the model. Computer models are therefore works of scientific art, not mathematical or statistical proofs.

This is a major reason that the issue of climate change is, and will likely remain, contentious, even within the scientific community. There is no way to prove that the models are based on the right assumptions and have captured all the relevant dynamics—and therefore to demonstrate conclusively that global warming is due to human activity, not natural cycles. A model that accurately mimics a system's behavior is not thereby verified: it could be coincidental. Even though the "scientific consensus" is that anthropogenic warming is the critical factor driving climate change, mere models can never compel belief—especially among those who have a vested interest in the economic status quo.

The goal of the scientists who study complexity is therefore more modest—to improve their understanding of the dynamics, patterns, and tendencies of complex adaptive systems, not to make predictions (as particle physicists can do, despite the seeming spookiness of quantum reality).

Among other things, in such systems small inputs can cause outputs that are disproportionate to the input by triggering a cascade of positive feedback. Or given enough time, small inputs can slowly amplify into large effects. Take the concatenation of consequences unleashed by the automobile. Beginning as a simple transportation alternative to the horse, it has radically transformed society—our built environments, our family structures, even our foreign policies. Conversely, large inputs can be decisively damped by strong negative feedback.

To put it another way, complex adaptive systems are both ultrastable and ultrasensitive at the same time. The challenge is to know which one we are dealing with. The journalistic cliché—that butterflies flapping their wings in Australia can cause torrential rains in the Andes and drought elsewhere—is an exaggeration that nevertheless captures the intrinsic unpredictability of nonlinear dynamics.

If the natural world is the consequence of a multitude of causes and conditions interacting in a nonlinear fashion over a very long time to produce a state of complexity that, although entirely lawful, can never be fully known—not in the present, much less in the future—then the modern hubris is belied. We shall never bend such a complex world to our simplistic purposes.

Indeed, at the heart of complexity lies something like an ancient mystery. The hidden order revealed by chaos theory is uncannily Platonic: invisible "strange attractors" inhabiting virtual "phase space" govern many complex phenomena. In effect, says Gleick, chaos theorists have rediscovered the Idea: "Behind the particular, visible shapes of matter must lie ghostly forms serving as invisible templates."[26] Faced with such mystery, respectful coexistence, not domineering control, will need to be our maxim going forward.

Chaos theory ties together all that we have discussed. First, the cosmic process is everywhere similar: the basic principles of order are universal, and self-organization reigns in nature from microcosm to macrocosm. Second, the fundamental character of that process is nonlinear; the mechanical is at best a special case, at worst an aberration. Third, what we observe as physical is actually the manifestation of something mental: "invisible templates" that create standing waves in the stream of space-time. It would seem that modern physics has indeed, just as Heisenberg said, "definitely decided for Plato."

The essential point is this: the new physics is fundamentally ecological. Everything is connected to everything else, and nothing exists in isolation because all phenomena are part of a larger, unified whole whose texture is determined not by the objects that it contains but by the complex tissue of interrelationships that create the contents. Thus the wisdom and ethic of ecology emerge equally from physics. Indeed, if even inert matter is intelligent to some degree and natural systems are more like beings than machines, then the case for humility, moderation, and connection grows stronger.

This conclusion is reinforced when we delve more deeply into certain specifics regarding physical limits, balance, and interrelationship. The place to begin is exponential growth.

Exponential growth is nothing more than compounding, as when interest accumulates in a savings account. Because new growth is added to previous growth, even a modest rate of increase will double the original amount in a relatively short time—for instance, a quantity growing at a steady 3.6 percent doubles in twenty years.

What is insidious about exponential growth is that it fulminates. The increments of growth are small at first, then large, then very large, and finally astronomical. Furthermore, the terminal explosion of growth comes with relatively little warning. After seven doublings, a quantity will be 128 times bigger, after fourteen 16,384 times bigger, after twenty-one 2,097,527 times bigger, but then after only seven more doublings it blows up to 268,483,456 times the original quantity. When graphed, an exponential growth curve therefore has the shape of a hockey stick. The upward trend is gradual at first, but as growth continues, the curve begins to bend more and more steeply toward the perpendicular, and it ends by rocketing to infinity.[27]

The practical import is that when a quantity is already quite large due to previous compounding, any further increase can be overwhelming—*because the next doubling will be approximately equal to all of the previous growth*. That is precisely our predicament with respect to carbon dioxide in the atmosphere. Having allowed emissions of this and other greenhouse gases to grow very large over the centuries since humankind began to depend on fossil fuels, we are now poised, in just a few decades, to release an amount of these gases equal to all that we have emitted hitherto, with consequences that are unpredictable but that could be dire. Yet it was only yesterday, comparatively speaking, that mainstream scientists began to understand that humankind had become a geophysical force capable of upsetting the global climate regime.[28]

The same insidious process works in reverse. Even a very large resource can be suddenly exhausted if the rate of exploitation grows constantly. That is our predicament with respect to petroleum After a century and a half of industrial development fueled primarily by oil, the world's appetite for petroleum products is now gargantuan, and global demand continues to increase as countries like China and India experience a rapid growth spurt. But many geologists believe we are quickly approaching, if we have not already passed, the halfway mark. That is, we have used up about half the earth's original endowment of exploitable reserves of liquid petroleum.

Exactly how much remains of our original endowment of hydrocarbons is controversial, but those in the best position to know do not disagree by orders of magnitude. In addition, there is near unanimity among them that the days of conventional oil and gas are numbered, so extracting the remaining resources will be much more difficult physically and expensive financially, as well as much more costly in terms of energy itself. In any event, a doubling of usage at this point would require us to extract in the next few decades an amount roughly equal to all that we have consumed up to now—in other words, virtually all of the remaining half, which is an impossibility owing to a host of geological, economic, and technical constraints that will limit the rate and drive up the price of extraction. We will experience serious economic, social, and environmental effects well before then. Again, despite warnings by various Cassandras over the years, a wider awareness of our predicament is only now dawning.

The point of this brief exposition of exponential growth is to make the physical limits of a finite planet concrete—something that merits our full attention here and now, not in some eventual future that may never arrive. As soon as one becomes aware of an approaching limit, it is already quite late. The momentum of growth has carried the ship into dangerous waters, with not much time or room left to maneuver.

The problem is that we are pushing the limits of the planet across the board. We are emitting not just carbon dioxide and other greenhouse gases but also a host of other harmful chemicals, compounds, and hormones. And we are depleting not only petroleum but also metals, materials, soil, and water—that is, most essential resources. In addition, having creamed off the easiest and best, we must now scrabble for the rest—for example, petroleum located under many fathoms of water and miles of sediment. The difficulty and cost of exploitation is fated to increase exponentially as we struggle to extract the remaining, lower-grade reserves of fuels, metals, and materials—and there will come a point where the game is no longer worth the candle because the costs of extraction equal or exceed the value of the resource.

The challenge of exponential growth is exacerbated by the laws of thermodynamics, which operate as internal or systemic checks on continued economic growth and technological progress. Matter and energy can be neither created nor destroyed, only transformed from one state or condition to another, and every transformation has a price. For example, when we burn coal to generate electricity, we get the heat that makes the steam that turns the turbine to generate the electricity but also various gases (among them carbon dioxide) as well as particulates, sulfuric acid, ash, and other unwelcome byproducts. From the physicist's point of view, the books balance—there is just as much matter and energy in the total system as before—but what remains is qualitatively poorer. So energetically rich matter is converted into relatively useless materials and into heat energy—some of which is used to do valuable work, but most of which is wasted. Even the electricity that does the work ultimately degrades into waste heat. The rich energy potential in the original lump of coal has been effectively exhausted. This loss of usefulness is called *entropy*.

Rather than engage in an extended technical exposition of the second law of thermodynamics, a homely analogy can be used to elucidate entropy.[29] Nature has a tax system more onerous and exacting than any that has existed since the late Roman empire. For example, when coal is burned in a power plant, only 30 to 40 percent of the potential energy is turned into electricity. The rest is dispersed into states or forms that are relatively useless or even noxious. In other words, nature taxes this particular transaction at a rate of 60 to 70 percent—a ratio that is typical.

Better technologies can reduce the tax but not by much once all the thermodynamic costs are tallied up: nature's tax rates are intrinsically high. To put it another way, nature is not very efficient (by human standards). For every gain that humanity makes by converting matter and energy into something it deems useful, it generates a corresponding thermodynamic loss that is typically around twice as large as the gain.

We are therefore trapped in a vicious circle. The more we produce and consume, the greater the increase in entropy—that is, the more depletion, decay, degradation, and disorder in the system as a whole, even if this brings us hedonistic or monetary benefits in the short term. That we have recently become aware of an array of environmental problems owing to escalating human demands on nature—all of which represent past-due bills from the thermodynamic taxman—is an indication that, once again, we have sailed into dangerous waters and are threatened with shipwreck.

The upshot is that we are leaving a time of relative thermodynamic abundance and entering a new era of thermodynamic scarcity that will resemble preindustrial times in important respects. Modern industrial civilization with all its appurtenances owes its existence to concentrated resources that are becoming increasingly scarce and expensive. They require ever-growing amounts of capital, energy, and expertise to exploit, a trend that cannot continue indefinitely. For instance, in the foreseeable future, well before fossil fuels are finally exhausted, the capital costs of exploitation will be exorbitant, the net energy yield minimal, and the environmental costs unbearable. When this happens, civilization will once again have to rely primarily on diffuse energy sources that lack the energy density needed to sustain an economy predicated on concentrated resources. For what matters is not the quantity of energy but its quality because this determines how much work it can do. This shift to lower-quality resources—primarily solar energy—has profound social and political implications. A civilization based on diffuse energy cannot have the same shape as one based on concentrated energy.

All this raises a question: if nature is so inefficient in engineering terms and has only the dispersed energy of sunlight for fuel, what accounts for its astonishing richness and diversity? As was shown in the previous chapter, limitation is a creative force that pushes nature to invest in quality, not quantity. The trend in ecosystems is from the relatively simple and wasteful pioneer stage characterized by competition toward the more complex and cooperative climax stage distinguished by mutualism.

To put it another way, nature internalizes thermodynamic costs by using the same matter and energy over and over to extract a maximum of life and complexity from a minimum of resources. Life, says James Lovelock, "is like a skilled accountant, never evading the payment of required tax but also never missing a loophole."[30] This exploitation of thermodynamic loopholes creates the rich tapestry of interrelationships characteristic of climax ecosystems. Nature is highly efficient in thermodynamic terms. The steady flow of solar energy is not simply consumed but is instead used to build up a rich capital stock.

Blinded by hubris, modern man does exactly the opposite. Like a spendthrift heir, he consumes the physical and biological capital of the earth. But you cannot eat your cake and have it too or spend capital and still enjoy income. Modern economies routinely ignore the critical distinction between stocks and flows in physical systems. Even when it would be possible to live on the income flow, we invade the capital stock. For example, we cut down whole forests for timber, fuel, and furniture instead of taking only what these complex ecosystems can comfortably yield over the long term. Or we vacuum up school after school of fish until the population crashes. And we ramp up agricultural production by mining the soil instead of nourishing it. If current trends were to continue (which they cannot), the human race would eventually appropriate so much of net terrestrial primary production that there would be nothing left but farmers' crops and the pests that infest them—an impossibility.[31]

The failure to distinguish flow from stock is particularly egregious with respect to the fossil capital accumulated over eons of geological time. Petroleum is the obvious example. Until very recently, we have treated the black gold deposited in the earth ages ago as if it were an infinite flow resource—when, in fact, it is a limited, precious, nonrenewable capital stock that is being rapidly depleted. Although awareness of our plight is growing, our entire way of life depends on consuming "the devil's excrement." We can scarcely imagine a future in which the flow of petroleum becomes a trickle and so have done almost nothing to prepare for that eventuality.

The case of water, the most critical resource of all, is even more alarming. Complex civilizations can exist without petroleum, but none can survive without water. And although water is the epitome of a flow resource, human demand has so far exceeded annual supply that almost everywhere in the world fossil water is being mined to support a burgeoning population that seeks to live ever higher on the ecological hog. This cannot continue.

In sum, a human economy based on money is colliding with a natural economy rooted in thermodynamics. Everywhere we turn, the human race is living beyond its means—disrupting or even destroying the natural systems that are the matrix of human life and depleting in generations the capital stocks accumulated over millennia or even eons. Our pursuit of economic wealth is impelling us toward thermodynamic bankruptcy. Again, the lesson is that seeking to maximize short-term human gain is self-destructive. Like ecology, physics ordains optimization as the only viable strategy for long-term survival and well-being. In other words, it ordains a climax civilization that prospers over the long term by internalizing thermodynamic costs to maintain a rich capital stock.

Enough has now been said to show that physics and ecology are fundamentally alike in the challenge they pose to human pretensions. Like the biosphere, the physical world is characterized by inescapable limits on human action and understanding. Ignoring these limits upsets preexisting natural balances and ultimately menaces the sustenance and survival of the human race. Humanity does not stand apart from the whole system. We exist because of the system, and our continued existence requires understanding and respecting the mutual interrelationship that binds man's fate to the rest of nature, living and nonliving alike. From this follows the same rebuke to human hubris and the same set of natural laws enjoined by ecology. Indeed, the case for humility, moderation, and connection becomes even more compelling.

With respect to humility, for example, what is now evident is not the power of the human mind but its weakness. A certain modesty is in order. What we do not know is vastly greater than what we do, what we think we know may well be wrong, and what we do know is only a parable. In other words, we should not behave as if we were the omniscient masters of the universe.

With respect to moderation, humanity can no longer treat the earth like a banquet at which it is free to gorge without regard for the consequences. As it is, we are trying to eat the fruit from a tree that we did not plant (and could never recreate) while simultaneously chopping it down for firewood. Human demand must therefore be moderated to a level that can be supported by the available flows, the capital stock that produces these flows must be preserved at all costs, and the remaining reservoirs of fossil capital should be conserved as much as possible. In other words, we cannot continue to behave as if we were the omnipotent lords and possessors of the universe.

Similarly, with respect to connection, we now know that the extent and depth of interrelationship in the cosmos are such that we exist in a kind of secret correspondence with all of creation. As Alfred North Whitehead expressed it, "any local agitation shakes the whole universe."[32] From such a realization follows an ethic eloquently articulated by Einstein. Calling man's feeling of separation from the rest of the universe "a kind of optical delusion of his consciousness," he drew an inescapable moral conclusion:

This delusion is a kind of prison for us, restricting us to our personal desires and to affection for a few persons nearest to us. Our task must be to free ourselves from this prison by widening our circle of compassion to embrace all living creatures and the whole of nature in its beauty.[33]

4
Psychology

Life has always seemed to me like a plant that lives on its rhizome. Its true life is invisible, hidden in the rhizome. The part that appears above ground lasts only a single summer. Then it withers away—an ephemeral apparition. When we think of the unending growth and decay of life and civilizations, we cannot escape the impression of absolute nullity. Yet I have never lost a sense of something that lives and endures underneath the eternal flux. What we see is the blossom, which passes. The rhizome remains.
—C. G. Jung[1]

The twentieth-century revolution in physics pioneered by Albert Einstein was paralleled by equally profound breakthroughs in psychology. The rediscovery and mapping of the depths of human consciousness initiated by Sigmund Freud and continued by his followers, especially by his estranged protégé Carl Gustav Jung, have been supported and amplified by the findings of numerous psychologists, anthropologists, ethologists, and neuroscientists. Although consciousness itself remains a mystery, we now understand both human nature and the limitations of the human mind better than ever before. As we have seen with respect to ecology and physics, however, this new understanding turns out to be not so very new after all. As Freud acknowledged, he retraced the pathways trodden by ancient poets.

The portrait of the psyche that emerges is cautionary. As much as contemporary humans would like to believe that we have transcended our evolutionary origins, our animal nature lives on within us—in our genes and in our minds. Witness the architecture of the human brain, in which the cerebral cortex enfolds a mammalian limbic system wrapped around a reptilian core. Hence, said Jung,

> Every civilized human being, however high his conscious development, is still an archaic man at the deeper levels of his psyche. Just as the human body connects us with the mammals and displays numerous vestiges of earlier evolutionary stages going back even to the reptilian age, so the human psyche is a product of evolution which, when followed back to its origins, shows countless archaic traits.[2]

In effect, Jung concludes, a "2,000,000-year-old man" dwells in all of us. Even the distinctively human part of our nature associated with the cortex is irredeemably Paleolithic.[3] As a consequence, men and women are constantly agitated by primordial drives and conflicting emotions that they only partly understand and struggle to control—and that they are usually not even aware of. Much is healthy and good in human beings, but we have propensities for sickness and evil that must not be ignored.

Anthropology supports this bleak assessment of the human psyche. With few exceptions, there are no harmless people, and the savage mind, whatever its virtues, is often prey to unconscious forces and raw emotions (and is therefore the author of savage behavior). A review of the anthropological literature reveals three seemingly universal tendencies of the human mind: we are prone to superstition and magical thinking, we are predisposed to paranoia, and we project our own hostility onto others.[4] In essence, says Melvin Konner, chronic fear pervades the psyche and drives human behavior.[5] Although the last word has yet to be spoken, there seems to be an emerging scientific consensus: we humans are a volatile mix of animal, primal, and civil—a tangle of emotions and drives that all but guarantees inner and outer conflicts.

That human nature is partly animal nature is not entirely a bad thing. Instinct is necessary for a healthy psyche and a moral society. But for human beings to live peacefully in crowded civilizations, the more bestial and savage aspects of man's nature have to be actively discouraged by society. Konner puts it more forcefully. Because of our fear-driven antisocial propensities, we humans are "evil" by nature and therefore need a "Torah," or an equivalent ethical code, to forestall the war of all against all.[6] In practice, this means that mores are essential because they tip the balance between good and evil in human nature. Good ones turn fallible, passionate men and women into reasonably upright members of society, while bad ones turn them into feral menaces to society.

This conclusion does not follow from theory alone; it has been empirically demonstrated. The social psychologist Stanley Milgram showed how simple it is to create little Adolf Eichmanns who obediently inflict severe pain on hapless experimental subjects.[7] In an even more frightening experiment, his colleague Philip Zimbardo contrived to convert ordinary, presumably decent students into punitive monsters. In the infamous Stanford prison experiment, student volunteers were randomly assigned to be either guards or prisoners. In a matter of days, the former turned harsh and sadistic, the latter cringing or rebellious, and the experiment had to be aborted to avert physical harm to the prisoners.[8]

In effect, psychology has rediscovered what were once called "the passions"—the welter of conflicting and potentially dangerous impulses and emotions that lurk in every human breast and that threaten to erupt under the slightest provocation unless they are kept in check by personal character or social control. Recall the words of Burke: "Society cannot exist unless a controlling power on will and appetite be placed somewhere." The choice is between self-imposed "moral chains" or externally imposed "fetters."

In his *Politics*, Aristotle identified the essential political challenge:

For as man is the best of the animals when perfected, so he is the worst of all when sundered from law and justice . . . [because he] is born possessing weapons for the use of wisdom and virtue, which it is possible to employ entirely for the opposite ends. Hence, when devoid of virtue man is the most unholy and savage of animals.[9]

When individuals gather in crowds, the challenge increases by orders of magnitude because fear, greed, and anger are contagious. As Gustave Le Bon pointed out long ago, crowds amplify every human defect and manifest many new ones of their own. "The masses," said Jung, "always incline to herd psychology, hence they are easily stampeded; and to mob psychology, hence their witless brutality and hysterical emotionalism."[10] Nietzsche was even more scathing: "Insanity in individuals is something rare—but in groups, parties, nations, and epochs it is the rule."[11]

The greatest weapon of mass destruction on the planet is therefore the collective human ego. History teaches that the human capacity for evil is virtually unlimited. Unless wisdom and virtue are deployed to counteract ego's potential for destruction, actual destruction is inevitable as men and women forget their better nature and become unholy and savage animals.

This new yet old understanding of human nature is enough by itself to demolish modern hubris. Infinite social progress is as much of a chimera as infinite material progress. The "2,000,000-year-old man" is what he is and will not be improved, only tamed. Indeed, at this point in human history, the essential task is forestalling racial suicide, not pursuing social perfection.

To this cautionary portrait of human nature, we must now add the limits of human cognition. As has been shown, the human perceptual apparatus is a trickster. We are in touch not with reality but with a kind of shadow play projected onto the screen of the psyche by invisible deep structures. We have also seen that even the finest intellects struggle to comprehend complex, self-organizing systems, for nature does not make it easy for us to know reality. But the fault does not lie in nature. The human mind was simply not created to unravel the mysteries of quantum mechanics or to comprehend the intricate dynamics of the global climate regime. It was instead cobbled together and then honed to perfection by evolution for one specific purpose—survival as hunter-gatherers on the African savannah. We are Jung's "2,000,000-year-old man" not just emotionally but also cognitively.

We are hardwired to perceive in certain ways and not in others. Above all, human cognition is "designed" for concrete perception, so primal peoples are masters of what anthropologist Claude Lévi-Strauss called "the sciences of the concrete."[12]

This is by no means an inferior mode of thought. The savage is not, as we tend to think, a mere captive of strange fancies and outlandish beliefs. He is actually more of an empiricist than the physicist because he perceives his world directly and immediately whereas the latter filters nature through an elaborate intellectual apparatus made up of mathematical, theoretical, and technological lenses. So the abstraction associated with literacy, civilization, and, above all, scientific investigation is not natural but acquired—and only with great difficulty after years of schooling.

Even schooling cannot entirely eradicate the innate propensity for concreteness in the human mind. For instance, we daily commit the epistemological sin of reification—regarding abstractions or ideas, such as energy or the market, as if they were somehow as real as rocks and trees rather than constructs that help us understand complex phenomena. Likewise, our opinions have a tendency to become "set in concrete," resisting all evidence to the contrary.[13] But perhaps the most egregious instance of what Whitehead called "the fallacy of misplaced concreteness" is that so many otherwise sane human beings believe in the absolute, literal truth of the manifestly mythological accounts contained in various scriptures—refusing to accept archeological and historical evidence to the contrary or even to entertain the possibility that these accounts could be fingers pointing at the ineffable rather than expressions of concrete truth.[14]

Sadly, many, if not most, human beings are not capable of rising very far above Piaget's concrete operational stage of cognition.[15] Hence they cannot be said truly to comprehend the social and physical reality of life in complex civilizations—a life far removed from the comparatively simple and concrete existence of the hunter-gatherer, which centered on day-to-day survival amid an intimate circle of kinsmen and friends.

As a corollary, the untutored human mind focuses on the present and the dramatic. The imperative of survival on the savannah made us sensitive to immediate or striking dangers—but comparatively oblivious to long-term trends, risks, and consequences, especially ones that are inconspicuous. Our attention is not grabbed by the creeping destruction of habitat, the imperceptible extinction of species, the continual accumulation of pollutants, the gradual loss of topsoil, the steady depletion of aquifers, and the like. Rather, we tend to fixate on dramatic symptoms (such as the occasional major oil spill) while ignoring the far greater long-term threat to ecosystems posed by quotidian events (such as the daily dribble of petrochemicals from a multiplicity of sources, which is far greater and much more damaging over the long term). Unfortunately, dribbles are not the stuff of melodrama and so tend not to register strongly, even when brought to our attention by the media. So it takes a crisis to thrust stealthy perils into full awareness.

Unfortunately, says biologist Richard Dawkins, the human brain was simply not built to understand slow, cumulative processes like evolutionary or ecological change, which demand an acute sensitivity to the long-term consequences of small changes.[16] Since long-term observation and planning were not critical for our early survival, these mental attributes were not reinforced by evolutionary selection. Ecology and its implications are therefore poorly understood, even by the informed public. More generally, the human mind's inability to escape the clutches of the present leads to the habitual, shortsighted pursuit of current advantage to the detriment of future well-being.

In addition, the survival imperative endowed us with a host of cognitive shortcuts—unconscious mental algorithms that may have been essential on the savannah but that must be consciously set aside if we humans are to live sanely in civilization. For example, the human mind tends to be quick to decide. Like any animal, we are emotionally wired for fight or flight, which means that our savage minds are also cognitively wired to jump to conclusions. When early humans spotted a tan shape lurking in the elephant grass, the minds that decided "lion" soonest had the best chance to pass their genes down to posterity.

The human mind is also dualistic, so it is constrained, if not compelled, to choose one pole or the other—fight *or* flight, black *or* white, right *or* wrong—not the middle ground. This has been experimentally demonstrated at the perceptual level: when humans look at a classical optical illusion, they see either the lady or the vase, never both at once. In other words, the human mind naturally dichotomizes, creating the common oppositions of "good" and "bad," "us" versus "them," the "two sides" of any issue, "left" against "right" in politics, and so on. Unfortunately, as F. Scott Fitzgerald noted, it takes a first-rate intelligence to hold two opposing ideas in mind at the same time and still continue to function, so untutored minds readily affix themselves to one of the poles and oppose the other. This explains the perennial conflict between believers and infidels that has occasioned untold historical misery.

The upshot is that we humans have multiple built-in mental weaknesses and limitations. Just as the ability to perform sustained analysis is not natural to the human mind and requires both talent and training, so the virtues of nuance, wisdom, prudence, and forethought are not innate and must be painstakingly developed through education.

What, then, is the forte of the human mind? Metaphor. The primary way in which the human mind makes sense of things is via analogy. The brain instantaneously compares incoming stimuli with its library of stored images (or sounds, smells, and so on) and "sees" that it is "like that" and is therefore a "lion" or whatever. In other words, we do not (and cannot) know what reality *is*. We can know only what it is *like*—that is, metaphorically. And this is so whether the metaphor is a verbal trope or a mathematical formula.[17] Goethe's aphorism perfectly summarizes this understanding: "All phenomena are merely metaphorical."[18]

It follows that even the scientific enterprise depends completely on analogy. As the physicist Brian Arthur says, "Non-scientists tend to think that science works by deduction. But actually science works mainly by metaphor."[19] Indeed, it must, for as physicist Robert Shaw observes, "You don't see something until you have the right metaphor to let you perceive it."[20]

Scientific metaphors are self-consciously formal models or theories. But these words reveal their metaphorical character. The word *model* means "likeness" by definition, and *theory* (which comes from the same root that gives us the word *theater*) is a kind of show in which the data are woven together to tell a plausible story about reality. In fact, as we have seen, the mathematical formulas that physicists use to parse nature are not reality itself but "parables"—in other words, stories, albeit extremely precise ones that allow for accurate predication. The task of science is therefore to find the metaphors that best stand for physical reality.

Human understanding is thus a process of lifting ourselves by our metaphorical bootstraps: we use what we already know to get at what we do not know by analogy. So all of human culture is a concatenation of metaphors until we arrive at those grand metaphors, such as Newton's clockwork universe, that shape whole eras. Bootstrapping being all but impossible in practice, however, we do not build the chain from the bottom up but from the top down. The image at the core of the grand metaphor governs the choice of metaphor and hence the process of understanding at lower levels. Contrary to the doctrines of empiricism, therefore, beyond the most basic (preconscious) levels of perception, our senses do not tell us what reality is: we instruct our senses in how to see the world. As Einstein noted, "It is the theory which decides what we can observe."[21]

In short, metaphors are the obligate lens through which men and women perceive and organize reality, cognitively and intellectually. As psychologist Julian Jaynes puts it, we are by necessity rough-and-ready poets: we construct a coherent world out of metaphorical images, but we do so haphazardly and unconsciously instead of self-consciously like writers and artists.[22] As Aristotle said in his *Poetics*, "the greatest thing by far is to have a command of metaphor [for] it is the mark of genius."[23]

As a consequence of this utter dependence on metaphor, human beings are adapted to understand life not as formula but as story: "The Universe is made of stories, not of atoms," said poet Muriel Rukeyser.[24] The human mind is virtually compelled to construct coherent narratives out of even the most fragmentary data. And it is not only children who crave stories: a vast entertainment industry thrives by providing them to the masses. Even at a more rarified intellectual level, it is remarkable that what we take to heart and remember is not dry exposition, no matter how cogent, but the striking image—in philosophy, Plato's cave; in literature, Dostoyevsky's Grand Inquisitor; in physics, Schrödinger's cat; in politics, Hobbes's state of nature ("solitary, poor, nasty, brutish, and short"); and so forth.

As these examples suggest, the cognitive process never occurs in a rationalistic vacuum. As obligate poets, we necessarily add feeling to understanding. The cerebral cortex simply has no way to think in isolation from the older subcortical layers of the brain that are emotionally attuned to symbolic meaning. So the human mind and heart crave myth and religion: a dry, abstract exposition of evolutionary theory can seem like thin gruel compared to the story of Genesis.

Even if one disdains the metaphysical *reality* of myth, one can hardly deny its psychological *function*. Science has neither abolished the need for human beings to make sense of their world—not merely intellectually but personally and morally as well—nor has it done away with their desire for satisfactory answers to the suffering attendant on the vicissitudes and ultimate finitude of life. As Wittgenstein, the philosopher who pushed the hyperrationalistic doctrines of logical positivism to their bitter end, put it, "We feel that even if *all possible* scientific questions be answered, the problems of life have still not been touched at all."[25] Beyond rational explanation lies a vast realm of silence—"Whereof one cannot speak, thereof one must be silent," said Wittgenstein—that contains all that is most important to us as human beings.[26] In short, without an emotionally satisfying story, the average man or woman simply has no answer to the riddle of life and death and is therefore liable to lapse into a state of spiritual vertigo.

This explains why ideology has such a tenacious grip on the modern mind. Lacking a coherent myth by which to live, human beings have desperately cast about for a substitute. Since it is unfortunately natural for the mind to fixate on *the* answer, on the *one* right way, human beings will tend to believe a story that purports to explain everything that is wrong with the world and that provides a simple solution—the overthrow of the bourgeois class, the abolition of sexual repression, the forced conversion of infidels, the war to end all wars, or the like. To put it in the terms made famous by Isaiah Berlin, the many are relatively simple-minded hedgehogs who thirst to know the one big thing; the few are sophisticated and nuanced foxes who know many things and who understand that life is not simple.

So the rationalists who consigned myth to the dustbin of history have been outflanked by the human mind, which so craves a story that it will swallow evil and absurdity rather than go without one. The refusal to deal constructively with the "irrational" simply leads to greater irrationality. The restriction of reason to a narrow ambit does not scotch superstition but instead leaves the major part of human life and experience outside the realm of correction by reason. Freud and Jung therefore provide the underlying dynamic for Le Bon's crowd behavior. Unless irrational forces within the individual are given a constructive cultural outlet, they will tend to go on a rampage, leading to mass manias, collective delusions, and religious frenzies.

This by no means exhausts the shortcomings of human cognition discovered by neuroscientists and evolutionary psychologists. But enough has been said to show that (1) we are constrained to see and comprehend the world in certain ways, (2) we therefore struggle to cope intellectually with the complex social and environmental conditions that we have created, and (3) we are prone to manias, delusions, and idées fixes that may be driven by underlying emotions but that are the also the consequence of defects in our thinking process. The gap between what human beings are cognitively capable of and the prevailing conditions in complex mass societies is therefore enormous.

To put it in another way, the human mind is far from a tabula rasa or simple stimulus-response machine. We are not rational automatons but living, breathing animals with deeply programmed predispositions and limitations. Our higher mental faculties—such as our ability to acquire language—depend on and are deeply influenced by a multiplicity of invisible deep structures. The upshot is that human beings are fundamentally constrained emotionally and cognitively. "Their passions forge their fetters," and what they can know of the world is a mere "parable" that owes as much (or more) to their passions as to their rational faculties.

Most of what has been said so far will be anathema to the hardcore rationalist or the social progressive. Even those less wedded to the Enlightenment dream of infinite progress may also feel that I am being excessively pessimistic. But the point is to be realistic about human nature and human capacities. If we pretend that people are better or more capable than they actually are and do nothing whatsoever to encourage them to be good beyond passing punitive laws, then moral entropy is inevitable. There will be a not so gentle slide toward increasingly antisocial behavior, and the key institutions of society will begin to break down. As parents know, their little bundle of joy is also a little barbarian who must be assiduously cajoled, if not directed, into becoming a reasonably civilized human being. This is even more so with each new cohort of barbarians in society at large. To reiterate, mores are all.

In addition, the situation is not as gloomy as it might seem. It may be that, as Immanuel Kant said, "From the crooked timber of humanity no straight thing was ever made,"[27] but we also have Aristotle's "weapons for the use of wisdom and virtue." We need to apply these weapons to compensate for our cognitive limits and foster our better nature. And we do have a better nature to foster. As shown by a number of recent studies, morality is intrinsic within the mammalian line of descent: animals too have a moral nature.[28] Even Konner concedes that human beings are not merely or irredeemably evil because a certain kind of ethics, found universally, is as natural to us as Euclidean geometry.[29] Compassion and other moral sentiments are therefore innate to human beings and need only be cultivated and supported for us to construct a social order that, however imperfect, is nevertheless reasonably harmonious and decent—and, perhaps more important at this point, conducive to long-term human survival.

The essence of what is necessary is captured in three Greek words—*therapeia*, *paideia*, *politeia*. Leaving the latter two for later, let us focus on the first. Depth psychology paints a cautionary portrait of the psyche, but it also offers a therapeutic path toward greater personal and social maturity—that is, toward wisdom and virtue, both in the individual and in society.

We have so far focused on the limitations of the human psyche and on the propensities toward excess, conflict, and imbalance that result from its evolutionary heritage. This understanding should incline us toward greater humility and moderation, both personally and collectively, but it is probably not enough. Ego does not want to be humble or moderate and therefore requires strong medicine to make it so. And for ego to swallow the medicine, it has to see clearly that humility and moderation are in its best interest—not just prudentially but because its own happiness and well-being are inextricably linked to the fate of all beings.

We have already seen the depth and extent of interrelationship within biophysical reality. Underlying the apparent diversity and separation at higher or individual levels is a fundamental unity at the deeper or collective levels. The same fundamental unity prevails in the psychic realm. Jeans echoes Einstein by saying that our apparently separate human consciousnesses "form ingredients of a single continuous stream of life."[30] In other words, our little lives are expressions of a much greater common life. The psyche is not an epiphenomenon but an autonomous, objective reality that has evolved biologically over the ages to contain structures and functions that are shared by all human beings. The cure for the little ego's defects is an awareness that we all participate in this common psyche and are therefore profoundly interrelated—in fact, united at the deepest levels of the psyche. In Jung's beautiful image, we are like flowers blooming from a single rhizome.

In what follows, I rely primarily on Jung and secondarily on Freud. Freud was the founding genius of depth psychology and produced striking metaphors that permeate modern thought. But Jung surpassed Freud in seeing that the sexual was only one aspect of the libido, however important, and in pushing the investigation of the psyche to greater depths than his notoriously dogmatic mentor was willing to go. (As Bruno Bettelheim makes plain, however, Freud was more "Jungian" than standard accounts of his person and work suggest.)

Freud can be seen as the last great figure of the Enlightenment. He viewed religion with a profoundly skeptical, even hostile eye. And he was in general loath to let go of the rationality exemplified by classical physics—even though, paradoxically, his own findings undermined that rationality. Jung, on the other hand, was a far more open-minded, post-Enlightenment thinker who enthusiastically embraced the new developments in biology and physics and who believed that myth and religion had essential functions in human life. (The charge of mysticism often hurled at Jung is largely unwarranted, however. For all his later dabbling in the seemingly occult, he was on the whole deeply empirical.)

Most important, however, Jung (along with Otto Rank) broke decisively with the millennia-old tradition powerfully reasserted by Freud—the so-called colonial view of the psyche, in which imperial reason vainly tries to subdue the unruly instincts and passions rather than coming to terms with them.[31] Like Euripedes, Jung and Rank understood that a rational dominion over the instincts was neither possible nor ultimately desirable.

Jung's rediscovery of the objective psyche is supported by numerous later studies—for example, the work of Konrad Lorenz and other ethologists.[32] Among its structures and functions are the emotional predispositions and cognitive propensities already discussed. But the objective psyche contains an even more important device for forming the contents of consciousness: the archetypes. These are, said Jung, "a priori structural forms of the stuff of consciousness" that determine the manner "in which things can be perceived and conceived" by human beings.[33]

The practical import is that each individual's psychic reality is constructed from universal patterns that reside in what Jung called "the collective unconscious"—the greater psyche that is common to all human beings as a result of their shared evolutionary heritage. These patterns shape, but do not determine, our psychic contents. In this respect, the archetypes are like Plato's Ideas. Just as there can be many variations on the idea of a chair, so the Form of the archetype can assume different forms in the lives of individuals. Every mother is different, but each expresses the Mother archetype, whether positively or negatively.

Jungian archetypes might seem to be spooky entities with no empirical foundation. However, as Jung pointed out, the physicist is in the same position as the psychologist because the underlying reality can only be inferred from its manifestations:

We must constantly bear in mind that what we mean by "archetype" is in itself irrepresentable, but has effects which make visualizations of it possible, namely, the archetypal images and ideas. We meet with a similar situation in physics: there the smallest particles are themselves irrepresentable but have effects from the nature of which we can build up a model.[34]

In other words, just as we can see the pattern of iron filings on a sheet of paper but not the magnetic force that causes it, so too we experience only the effects of the archetypal force and not the underlying archetype itself, which remains forever hidden and inaccessible.

In fact, universal patterns tantamount to archetypes govern not only the psyche but all of human life. There is a "universal grammar" underlying the multiplicity of human languages—such that, even if all existing languages were to vanish, the next generation would soon invent a pidgin that would in time evolve into a full-blown literary language. So too is there a universal pattern for human life underlying the multiplicity of different cultures.[35] For the anthropologist, human life is but a set of variations on universal themes, ranging from age gradients to xenophobia, that are aspects of this universal pattern.[36] Indeed, says Robin Fox, such is the force of the basic template for human life that should there ever be a new Adam and Eve, "they would eventually produce a society that would replicate ours in all essential features."[37]

If all this seems excessively "mythological," so be it. Recall that humans are at bottom irredeemably Paleolithic. Civilization is a thin veneer on a fundamentally primitive and archaic development. As we have also seen, metaphor is the basis of human understanding—so symbol and story, not rational discourse, are the mother tongue of the psyche. It follows that the human mind is essentially mythological, both in its operations and in its contents, and myth is the indispensable bridge between our inner and outer realities. Myths are not fantastic attempts by ignorant savages to explain reality in the absence of reliable methods to determine empirical truth. Rather, they are allegorical attempts, embellished with admittedly fantastic details, to express the interrelationship of the natural, cultural, and psychic aspects of reality—that is, the cosmic truth of kinship. In other words, the prescientific mythmakers did with imagination what the scientists have now done with theory—

describe the cosmos as a living being and work of art, not a lifeless automaton.

Any approach to the human psyche, however empirical, must necessarily be mythological. Our health depends on having a good relationship with the archetypal level, either through participation in a living cultural tradition that articulates an appropriate myth or through direct contact with the source of myth in the collective unconscious. Failure to be connected to the mythological depths of the psyche leads to disease—mental illness in the form of neurosis or worse, physical illness in the form of psychosomatic ailments, and social illness expressed as addiction, anomie, anxiety, and the like.

That psychosomatic illness arises from conflicts between the layers of the psyche has been confirmed by medical researchers. Neurologist A. T. W. Simeons, for example, says that our rapid cultural evolution has left us "cortex ridden"— that is, psychologically dominated by an upstart reasoning brain at the expense of the mammalian midbrain and the reptilian brain stem. But the human subcortex knows nothing and understands less of civilization. It therefore reacts to events as if we were still living as just another animal on the African plains, whereas the cortex is acculturated to modern life and responds accordingly. Civilized human beings are therefore torn between two conflicting realities. The psychosomatic illness that pervades modern societies has a real physiological basis; it is not just "all in the mind." Given the neurological facts of life, says Simeons, there is only one possible strategy for genuine physical health—to raise subcortical activity to conscious awareness and give it its just due, which is exactly what depth psychologists prescribe for mental health.[38]

Despite their seemingly mythological character, the archetypes are real and must be taken into account. They are psychological templates analogous to those found in the physical world and serve the same purpose—to organize psychic reality to accord with the limited capacities of the human nervous system. Although invisible, they manifest concretely in an individual's personal unconscious as "complexes." These in turn mediate and control our conscious thoughts, emotions, and actions in ways of which we are normally almost entirely unaware.

This lack of awareness is a problem. We unconsciously and neurotically act out our inner conflicts in ways that harm self and other. We therefore need to make the unconscious conscious. By becoming aware of the wellsprings of our motives and actions, we can begin to bring our life under some semblance of conscious control. As Freud said, "Where id was, there ego shall be." However, Freud's view of the human condition as expressed in *Civilization and Its Discontents* was decidedly pessimistic. Like a beleaguered garrison barely holding out against the immensity of the unconscious, ego's position was precarious. It was not even master in its own house. Moreover, the conditions of civilized life made repression and its painful consequences inescapable, so the best that psychoanalysis could hope to achieve was the exchange of neurotic misery for ordinary unhappiness. Thus, Freud called for heroic resignation in the face of life's adversity.

For Jung, however, shining the light of consciousness on what he termed "the shadow" was but a first step toward a genuine cure for the "sick animal" created by civilization. Chronic neurotic discontent dissolves only when the little ego establishes a relationship to something greater than itself—a transcendent "Self" that, through its connection to the collective unconscious, comes to a direct knowledge of its kinship with all being and attains some sense of the "numinous." So the Jungian cure is ultimately spiritual, not rational: "The decisive question for man is: Is he related to something infinite or not?"[39]

The goal of Jung's therapy was to have the patient go through a process that he called "individuation" to activate this greater, transcendent Self—not by ascending upward toward the light but by descending into the dark holding the torch of consciousness with the aim of harmonizing spirit and instinct. This deliberate, conscious integration of what is normally rejected and repressed because it seems negative or destructive allows the psyche to become whole. It is no longer torn between the demands of nature and culture.

Unlike Freud, for whom nature was an object of dread to be subdued by the rational ego, Jung found that the human animal was sick precisely because he had been uprooted from the soil of instinct by an excessively rational civilization that systematically frustrated his archetypal needs: "In the final analysis, most of our difficulties come from losing contact with our instincts, with the age-old forgotten wisdom stored up in us."[40] It was as if we were confined to one or two claustrophobic rooms in an ancient mansion filled with extraordinary riches: "Overvalued reason has this in common with political absolutism: under its dominion the individual is pauperized."[41] In effect, said Jung, modern men and women are in the same sad state as a primitive tribe dispossessed of its ancient folkways—utterly dispirited.[42]

What is worse, losing contact with the instincts turns man into "the devil's marionette":

The dammed-up instinctive forces in civilized man are immensely destructive and far more dangerous than the instincts of the primitive, who in a modest degree is constantly living out his negative instincts. Consequently no war of the historical past can rival in grandiose horror the wars of civilized nations.[43]

Paradoxically, the path to the numinous realm of the transcendent Self lies through the instincts—through coming to terms with our animal nature. The psyche contains all the opposites, so spirit and instinct are inseparable: one cannot be had without the other. More generally, the goal of therapy must be the integration of all the conflicting tendencies and drives of the psyche because repression provokes an inevitable and obsessive psychic demand for what has been repressed. Another way of stating the goal of Jungian therapy is that it aims at unifying the opposites in a "sacred marriage" that achieves psychic balance and wholeness and thereby true mental health.

Based on his own life and on his long clinical experience, Jung believed that he had found a way in which individuals could transcend their egoism without losing their autonomy along the way. Traversing the path of individuation, the patient first strengthens ego by integrating the instincts but then goes on to dissolve ego's rigid shell by reconnecting to the "something that lives and endures underneath the eternal flux." Out of this experience, which can be more or less powerful and complete, there arises an inner happiness relatively independent of external conditions, as well as a willingness to live a reasonably virtuous life in harmony with one's fellows. As Jung put it, "the natural process of individuation brings to birth a consciousness of human community precisely because it makes us aware of the unconscious, which unites and is common to all mankind."[44] Hence, "individuation does not lead to isolation, but to an intenser and more universal collective solidarity."[45] In essence, individuation means achieving the fullest measure of individual*ity*—but without the selfishness, competitiveness, resentment, alienation, loneliness, and narcissism attached to the individual*ism* of today. So individuation is "ultimately a matter of ethics": it makes us both truly human and genuinely civilized.[46]

To put the issue in a larger (and more Freudian) context, our excessively abstract, overly rational, and aridly mechanistic modern way of life explicitly repudiates Eros, the force that creates and sustains life in all its variety and loveliness and that also unites life forms into communities. However, since Eros inspires most of the things that make life worth living—beauty, joy, creativity, friendship, and love—to spurn Eros is to impoverish oneself or, even worse, to invite Thanatos, the opposing force that extinguishes life, severs connection, effaces beauty, and kills joy. Here is how Jung put it:

Eros is a questionable fellow and will always remain so. . . . He belongs on one side to man's primordial animal nature which will endure as long as man has an animal body. On the other side he is related to the highest forms of the spirit. But he thrives only when spirit and instinct are in right harmony. If one or the other aspect is lacking . . . the result is injury or at least a lopsidedness that may easily veer towards the pathological. Too much of the animal distorts the civilized man, too much civilization makes sick animals.[47]

The consequences of repudiating Eros are therefore profound. The conventional wisdom notwithstanding, civilization is endangered not so much by the natural and erotic per se but by the denial or repression of our instinctual nature. This turns human beings into sick animals whose frustrated drives and unconscious demands will be expressed pathologically in the famous "return of the repressed," which can take both individual and collective forms.[48] As Erich Neumann put it, "splitting off of the unconscious" activates dangerous impulses that "devastate the autocratic world of the ego with transpersonal invasions, collective epidemics, and mass psychoses."[49] In other words, the real menace to civilization is not our all-too-human nature but the barbarism that arises when an excess of civilization or the wrong kind of civilization thwarts or twists that nature and thereby creates a host of neurotics (and not a few psychotics).

Jung's individuation is therefore much more than an individual cure: it is also a genuine *therapeia*. It provides an avenue for thinking about how we might begin to create a reasonably sane and humane society out of "the crooked timber of humanity"—a society in which there is a rough harmony of instinct and spirit and in which all aspects of the human personality can flourish and yet be contained.

This containment is essential. Individuation has nothing to do with the amoral acting out of instinctual drives. The primordial error of the modern age was to think that the passions—the powerful forces of the unconscious—could be unleashed without causing harm. But repressing them is no better, for this creates a host of sick animals and sets up the return of the repressed. Ergo, the instincts must be consciously sublimated—that is, neither grossly expressed nor harshly repressed but instead transmuted by directing them to higher ends, both personal and social.

Jung was under no illusions about what might be necessary: "Mankind is, in essentials, still in the state of childhood—a stage that cannot be skipped. The vast majority needs authority, guidance, law. This fact cannot be overlooked."[50] Thus, although he was in general contemptuous of collectivities, Jung nevertheless sided with Dostoyevsky's Grand Inquisitor:

Collective identities are crutches for the lame, shields for the timid, beds for the lazy, nurseries for the irresponsible; but they are equally shelters for the poor and weak, a home port for the shipwrecked, the bosom of a family for orphans, a land of promise for disillusioned vagrants and weary pilgrims, a herd and a safe fold for lost sheep, and a mother providing nourishment and growth. It is therefore wrong to regard this intermediary stage as a trap; on the contrary, for a long time to come it will represent the only possible form of existence for the individual.[51]

So individuation is not a purely individual matter and cannot occur in a social vacuum. The unconscious holds formidable archetypal energies that must be personally sublimated and culturally channeled lest they assume dangerous forms. Making greater self-awareness (along with greater cognitive sophistication) a social reality, if not the overarching goal of human life, therefore presumes the existence of a social and political framework capable of modulating the psychic impulses of a multitude of individuals for the mutual benefit of all—in other words, a rich culture, a strong community, and a government genuinely devoted to this end. In short, individuation entails a real *politics* of consciousness.

What such a politics might be is the subject of future chapters. But given the consistent pattern across the three disciplines examined thus far, we start with a clear ethical basis contained in natural laws that emerge from the way things are. We need to be humble about our limitations, both external and internal; we need to moderate our behavior to avoid imbalance, both physical and mental; and we need to honor our intimate connection to all of life, both material and spiritual.

We also start with a clear intellectual basis, for psychology has, along with ecology and physics, "definitely decided for Plato." What are the archetypes but Platonic Ideas? Even if one dismisses the archetypes—or all of Jung—the evidence for the war between reason and passion within the human breast and, above all, for the existence of deep structures governing human perception would seem to be incontrovertible. In studying nature, it turns out that we are really investigating our own consciousness, for it is the nature of our own mind that decides what we can observe. The light turns into shadows on the wall of the cave not because there are Ideas out there in epistemological heaven but because there are hidden templates for organizing experience within the human brain and mind. As William James put it, "Reality is apperception itself."[52]

Even apart from their shared idealism, Jung's "psychology" and Plato's "philosophy"—or Jung's *therapeia* and Plato's *paideia*—have everything in common. Because the guiding truth for man is to be found in the light outside the imprisoning cave of the personal psyche, the task of every individual is to try to escape from this prison by accessing the primordial intelligence that underlies and animates both the phenomenal and psychic worlds. And the role of society is to support human beings in this endeavor.

The shade of Socrates must be rejoicing in Hades to learn that modern science, in seeking to discover a hard-and-fast physical reality free from all unseen forces, has only succeeded in reconstructing his shadowy cave. From this development follows philosophy in the Socratic mode—namely, a rejection of merely material values and a quest for wisdom and virtue. In no other way can we tame the passions of the "2,000,000-year-old man" so that he is fit for civilization and at the same time remake civilization so that it is fit for him.

The Politics of Consciousness

5
Paideia

> Mere purposive rationality unaided by such phenomena as art, religion, dream, and the like, is necessarily pathogenic and destructive of life. . . .
>
> Our loss of the sense of aesthetic unity was, quite simply, an epistemological mistake . . . more serious than all those minor insanities that characterize those older epistemologies which agreed upon the fundamental unity.
>
> —Gregory Bateson[1]

Paideia is *therapeia* writ large. *Therapeia* is the restoration of unity to the psyche, so that the archetypal needs of the "2,000,000-year-old man" are reconciled with the demands of civilized life. *Paideia* is the recovery of the "aesthetic unity" destroyed by a "mere purposive rationality." It is a cure for the major insanity of a way of thinking and living that ignores the "fundamental unity" and is therefore "necessarily pathogenic and destructive of life."

To put it another way, *therapeia* is about achieving instinctual and emotional sanity, and *paideia* is about achieving cognitive and intellectual sanity: what epistemology, what way of thinking, what worldview, what myth or metaphor will foster a sane, humane, and ecologically viable way of life over the centuries to come?

This is not a utopian quest to straighten the crooked timber of humanity. The passions can only be tamed, not transmuted, and politics as an arena of conflict will be with us forever. But greater sanity, both individual and collective, is within our grasp, provided that we ground ourselves in Bateson's aesthetic unity. With this as our premise, we can discover and then institute politically an Aristotelian rule of life that inclines our civilization toward a wise virtue instead of an unholy savagery. Unless we soon transform our purely instrumental relationship to life, we confront a bleak future of neurotic misery, social decline, spiritual poverty, and ecological destitution.

Let us distinguish *paideia* from education as we commonly understand it. Most of what we call education is in fact schooling in the methods and values of "mere purposive rationality." With rare exceptions, students attend universities not to become better or more universal human beings but to obtain a credential that will lead to gainful employment in the military-industrial-financial-political-media complex—or in the university itself, which has (sad to say) become a kind of knowledge factory allied to the complex.

So most students receive a narrow, technical education that prepares them for careers by rendering them marginally competent in mathematics or metaphor but rarely in both. Contrary to the assertion of many that we have entered a new golden age of knowledge and learning, our educational factories therefore "produce" graduates who are, by the standards of the past, profoundly ignorant. Even the so-called best and brightest of today are mere specialists who, as it is said, know more and more about less and less. Paradoxically, for all the resources devoted to the education industry, its "output" is a kind of idiocy.

Premodern thinkers like Montaigne and early modern thinkers like Locke had small personal libraries of a few hundred volumes, but they possessed a mastery of the Western tradition since earliest times and an integrated understanding of the workings of their respective societies. Whatever their personal or intellectual shortcomings, they therefore commanded an overview of the human condition that was close to the best that their respective ages could achieve.

We are intellectual paupers by comparison. An analog of Gresham's law in the economy of mind has devalued knowledge, first by swamping it in information and then by submerging it in data. We have an exact count of all the cats in Zanzibar, but meaning, context, and coherence utterly escape us.

By contrast, *paideia* is "the process of educating man into his true form, the real and genuine human nature."[2] It is education for excellence, education for statesmanship, education devoted to the Greek ideals of the beautiful and the good, education designed to inculcate a deep appreciation of the aesthetic unity of the world.

The aristocratic connotations are obvious. *Paideia* is intended to form an elite in its original, nonpejorative sense—"the best or most skilled members of a given social group." By seeking to develop a natural aristocracy instead of an artificial meritocracy, *paideia* directly contradicts the prevailing vulgar democratic ethos, according to which there can be no "betters."

But elites are inevitable. In any field of human endeavor, the cream rises to the top unless it is forcibly prevented from doing so. For instance, there are elite athletes and elite scholars. The realm of politics is more complicated yet not essentially different. Once the polity has grown too large to gather under an oak tree on the village commons, the burden and opportunity of governance necessarily fall on a relatively small group of insiders—that is, an oligarchy. The question is, is this oligarchy a genuine elite that governs according to an inspiring ideal and with some sense of noblesse oblige? Or is it a pseudo-elite—"a narrow and powerful clique"—that rules in its own interest and runs the state as a racket? Our current ruling class—a meritocratic oligarchy allied to a predatory plutocracy—seems to resemble the latter more closely than the former. Be that as it may, by the very nature of things, society and polity will always be directed by an oligarchy—but we must control the nature of the resulting ruling class to avoid winding up as oppressed subjects of a gangster elite.[3]

Elites are also necessary, for the reason given by Gustave Le Bon:

Civilisations as yet have only been created and directed by a small intellectual aristocracy, never by crowds. Crowds are only powerful for destruction. Their rule is always tantamount to a barbarian phase. A civilisation involves fixed rules, discipline, a passing from the instinctive to the rational state, forethought for the future, an elevated degree of culture—all of them conditions that crowds, left to themselves, have invariably shown themselves incapable of realizing.[4]

What is worse, as Jung points out, the absence of a good elite opens the door to a bad one:

The levelling down of the masses through suppression of the aristocratic or hierarchical structure natural to a community is bound, sooner or later, to lead to disaster. For, when everything outstanding is levelled down, the signposts are lost, and the longing to be led becomes an urgent necessity.[5]

In short, without a genuine elite—one capable of upholding a rule of life that fosters some reasonable degree of wisdom and virtue and that responds realistically and effectively to challenges—no society or civilization can long endure. *Paideia* is about discovering such an ethos and forming such an elite.

I propose the epistemology, ontology, and ethic articulated in the previous three chapters as the new ethos. This radical transformation of our way of thinking and being would ground civilization on ecological, physical, and psychological reality rather than hubris. It would also inspire us to make a civilization more worthy of the name.

The quintessential task of the new *paideia* will be to restore beauty to its rightful place in the pantheon of human values. It might seem strange to posit a lack of beauty as the most critical defect of a utilitarian civilization. Why is beauty important? And can it be said to be missing from modern civilization when all manner of so-called art is being produced? To answer the latter question first, James Buchan describes the plight of beauty in a totally commercialized culture:

The sensation of beauty cannot survive in the age of money: for any beauty must by exploited, reproduced a million times by every medium open to commercial ingenuity, till one can only cover one's eyes and stop one's ears. The sole aesthetic sensation of modernity is nausea: permanent, lethal nausea.[6]

In addition, when the arts flourish only at the margin of society, they fragment into subcultures populated by small coteries of specialists and aficionados. They no longer function as a school for human understanding as they did in the past. We have, for instance, nothing remotely comparable to the great dramatists of ancient Athens or Elizabethan England to help us grapple with deep cultural questions.

More important, what is the quality of the art being produced, especially at the so-called cutting edge? Does it inspire us with higher ideals or nobler visions of life? Does it connect us to our archetypal depths or to the wonder of creation? Or does it simply reflect back to us an increasingly disjointed, frenetic, and artificial way of life? Just as a plethora of man-made laws can never compensate for a lack of moral law, so too "art" that is not truly or essentially beautiful is a symptom of corruption, not health.

However beauty is defined, it is critical. First, the conspicuous absence of beauty from modern civilization has grave consequences. Beauty, says psychologist James Hillman, is "the Great Repressed, the taboo that never is mentioned," whereas molestation, addiction, perversion, and all the other objects of repression a century ago are now front-page news.[7] In consequence, says Hillman, the main locus of psychic oppression today has shifted from inside to outside—from the inner hell of personal repression to the outer chaos of a crowded, bureaucratic, and polluted way of life filled with stuff that is "nasty, brutish, ugly, cheap, shoddy, vicious."[8] But living in such a poisonous setting cannot be healthy. Ugliness numbs and sickens the psyche, rendering it increasingly dead to the world. For Hillman, "Beauty is an epistemological necessity; *aistheis* is how we know the world."[9] It is also "an *ontological* necessity": it uplifts the human spirit and reconnects us to the aesthetic unity that is the indispensable ground for individual and collective sanity.[10]

Second, if we are to reconcile the "2,000,000-year-old man" to civilized life, then we must make it worth his while by providing him with an environment that responds to his archetypal needs. The human subcortex, the repository of our animal heritage, craves the rich sensual stimulation of the natural world—that is, the kind of experiences for which the beauty-starved denizens of an artificial world willingly pay thousands of dollars to enjoy on vacation. There can be no *therapeia* without a *paideia* that elevates beauty into a chief feature of the civilization instead of something to be interred in museums or "collected" by wealthy eccentrics.

Finally, the poet John Keats's famous equation of beauty and truth is itself both true and beautiful. Beauty is indeed the deepest truth—a truth that is immediately and directly perceived as such, a truth that simply *is* beyond logic, ideology, or argument. And beauty is evidence for truth, whether in a painted portrait or a scientific equation. As physicist Steven Weinberg says, "We would not accept any theory as final unless it were beautiful."[11] So truth and beauty are psychically inseparable, which means that a world awash in ugliness is bound to seem false and unreal as well.

But what is this beauty that is so important? It is not fashion or style but something deeper—something that delights both the human subcortex and the right half of the cortex that operates in the artistic rather than analytic mode. In other words, beauty is what instinctively gladdens the heart of the "2,000,000-year-old man." A trained musician's appreciation of, say, a Beethoven sonata will be more refined and complex than that of the average listener, but the essence of the human aesthetic response now is no different than it was when our archaic ancestors danced and sang around the campfire or decorated their bodies with ochre and oil.

Beauty is not merely "in the eye of the beholder." We may not be able to define or measure it with scientific precision, but we can know it nonetheless. After all, we are able to experience the essential beauty of art produced by many cultures. The instincts are common to all of humankind, so there are universal patterns of aesthetic response.

The architect Christopher Alexander calls the aggregate of these patterns "the timeless way"—timeless because it is rooted in instincts that have not changed in forty thousand years.[12] And this way can be specified. Seeking to discover what makes towns and buildings beautiful, Alexander and his colleagues identified 253 universally found patterns that constitute a "pattern language" by which architectural beauty can be assessed and created.[13] It is no accident that Italian hill towns are almost universally considered to be architecturally beautiful: they exemplify this pattern language to a high degree.[14]

So although beauty is in one sense immeasurable, it has an empirical basis in human neurology and psychology and can therefore be consciously comprehended and created. Indeed, it must be, for unless we reawaken our anesthetized sense of beauty, we shall never appreciate the aesthetic unity of the cosmos or steer civilization away from its present trajectory toward self-destruction.

It follows that one major thrust of the new *paideia* must be to redress the radical imbalance in modern culture and education. The problem is that an instrumentally efficient civilization has exalted the values of the rational-logical left brain over those of the artistic-intuitive right brain.[15] To put the issue in the terms made famous by Pascal, we have overemphasized "reason," the faculty associated with our left brain, while slighting or ignoring the "reasons of the heart" that have their neurological seat in the right brain. The remedy for this defect is an aesthetic education in a broad sense that goes beyond mere music or art appreciation.

Two important reasons for this remedy have already been given. First, because art is the primary carrier of beauty within a culture, an aesthetic education is indispensable. Second, the metaphors and symbols carried by art are also the language of the deeper psyche, so if we wish to heal the split between "overvalued reason" and the primal passions, we must once again become fluent in that language.

Equally important, however, is the vital role that art plays within secular education. First, as Konner points out, a deep understanding of human nature (as revealed by behavioral biology and depth psychology) is rarely found in the writings of social scientists, whose shallow and Panglossian practitioners seem to believe that, with just enough social tinkering, we shall soon attain utopia. Instead, says Konner, it is the "artists of the soul"—poets, tragedians, and novelists—who most clearly reveal the truth of the human condition and who best instruct us in how to meet its challenges.[16] Indeed, Thomas considers artists to be the scientists of quality, for they illuminate realms that scientists beholden to mere quantity cannot reach.[17] Even those whose primary interests lie in physics or sociology rather than belles lettres should therefore study literature.

The second reason that secular education needs to include the arts was addressed in the previous chapter: human understanding depends utterly on metaphor. The more metaphors we have at our command, the more likely we are to come up with the appropriate template for understanding reality here and now. Even scientific discovery depends critically on metaphor. To put it the other way round, if a an abundance of metaphors is, as Aristotle said, the sign of genius and the font of creativity, then a paucity of metaphors due to an excessively narrow education is a prescription for intellectual stagnation and cultural idiocy.

Third, an aesthetic education is essential to develop the faculty of judgment, which we need both for understanding and for action. As physicist Roger Penrose points out, mathematical and scientific truths are determined not by logic alone but by a careful weighing of evidence that is driven, at least in part, by aesthetic criteria.[18] Einstein always maintained that intuition, curiosity, and imagination, not his thinking mind, had led him to his groundbreaking discoveries: "It was not my rational consciousness that brought me to an understanding of the fundamental laws of the universe."[19] In other words, even in science, the challenge is to find the *fitting* metaphor—a matter of judgment. In less rigorous fields, it takes real discernment to sense patterns hidden among a multiplicity of variables. In short, reality does not impose itself: we have to figure it out—judiciously.

More important, however, we need judgment to guide us through the thicket of human affairs. Lacking an aesthetic appreciation of context, texture, depth, perspective, nuance, and the like—that is, the whole range of meanings and possibilities—we are bound to blunder. As Whitehead noted, exalting the scientific over the aesthetic was a "disastrous error" that left us caught "between the gross specialized values of the mere practical man, and the thin specialized values of the mere scholar"—that is, between Wall Street and the Ivory Tower—without a solid or realistic basis for making critical decisions.[20] Nowhere is the faculty of judgment more necessary than in politics.[21] Friedrich Schiller made the connection between politics and aesthetics explicit. After witnessing the euphoria of the rights of man terminate in a reign of terror, he concluded that "Man will never solve the problem of politics except through the problem of the aesthetic, for it is only through beauty that man makes his way to freedom."[22]

To bring the matter down to earth, a system of education that prepares people for "careers" rather than for life in all its dimensions, especially its tragic dimension, will readily lead its graduates into Thoreau's trap of "improved means to an unimproved end."[23] We will unthinkingly build bigger and supposedly better mousetraps that do little or nothing to improve the quality of our lives or that even make them worse. Without judgment—the ability to assess the nature of the reality that confronts us, to conceive the likely consequences of alternative courses of action, and then to weigh the delicate balance between ends and means—we are little more than skilled barbarians whose shortsighted and imprudent decisions will terminate in ruin.

The essential thrust of the new *paideia* is a movement from quantity to quality and from matter to spirit. Our root problem is that we are approaching, if we have not already exceeded, the limits of material development. To shift our focus from outer expansion to inner cultivation, we need to develop Pascal's "heart," an organ of apperception and evaluation that has to be conscientiously trained if we are to develop its faculties to the fullest. Appreciation, intuition, discernment, imagination, vision, creativity, and the like—the qualities we will need for a rich, vibrant, successful culture going forward—exist to some degree in every human being. But a society dedicated to mere quantity has neglected to educate for these qualities, leaving us without the resources necessary to reconstitute a civilization that, as Jung said, has "sold its soul for a mass of disconnected facts."[24]

We require something like an expanded, more Platonic version of the classical liberal arts curriculum—more Platonic in that it would include music, gymnastic, and poetry in addition to logic, mathematics, and other rational disciplines. (Modern readers too often anachronistically mistake Plato's reason for a dominating instrumental rationality, when it is more like Pascal's reasons of the heart—that is, a faculty that can be developed only by "educating man into his true form.") Such a curriculum would accomplish the five essential purposes of an aesthetic education: it would foster beauty, cultivate the reasoning heart, build up our stock of metaphors, liberate intuition and imagination, and also teach realism and judgment. Unlike a merely technical or academic education, a truly classical education would enrich us physically, psychologically, emotionally, and intellectually. Above all, it would provide a powerful inoculation against the greatest of all delusions—that one's own view of reality is the only one possible.

The next critical element is the reconstitution of reason itself. Civilization needs a new heart and also a new mind. Although the aim of an aesthetic education is not to exalt poetry over physics but only to restore a healthy balance between the two, this will nevertheless require a very different mode of reasoning than that of the mechanical age—namely, the wider adoption of the ecological worldview (also called the *systems paradigm*) as the basis for thought and action.

Toward this end, a radical demystification of what we call reason is unavoidable. Every system, from shamanism to science, has its own principle of reason. Reason therefore resembles language. Just as language can never be objective or neutral—for all languages, without exception (mathematics included), express a particular worldview—so too reason is not an independent criterion to which one can appeal in favor of one system rather than another. The instrumental rationality that most moderns believe is the only valid form of reason was not forced on us by nature. It was espoused—in part consciously but in larger part unconsciously—to further the ends of domination inherent in the cultural values of Western civilization.[25] Now that domination is no longer an achievable goal, we need ecological reason.

Our thinking must become as complex as we now know nature to be. The aim is not to bend nature to our will—this is ultimately not possible—but rather to adapt ourselves to the way complex, self-organizing systems actually work. As Francis Bacon famously said at the outset of the scientific revolution, "For we cannot command nature except by obeying her."[26] We must, paradoxically, control ourselves in order to command nature—by accepting limits, preserving balance, and respecting interrelationship. What greater humility, moderation, and connection will mean in practice is best understood through examples.

Contrast two views of an old-growth forest. To economic man, the forest is a resource to be exploited "rationally" to achieve the greatest return on capital. This is best accomplished by cutting down all the trees, selling the lumber, and then either buying another resource to exploit in similar fashion or investing the proceeds to produce income (in some cases by replanting the devastated land with tree crops, which produces a counterfeit forest). There is little incentive to preserve the original capital stock by logging the forest more selectively over a longer time period because sustained-yield forestry entails a productive cycle of many decades. Since "a bird in the hand is worth two in the bush," the value of anything far in the future is "discounted" to zero by a "rational" economic actor.

This purely economic view of the forest is extremely narrow: nothing matters except maximizing profit here and now. By contrast, ecological man is all-inclusive. He takes into account all the elements ignored by economic man—not only the long-term practical benefits provided by living forests (flood control, air quality, soil protection, climate moderation, wildlife preservation, recreation, and so on) but also the larger social and political issues connected to the exploitation of nature and even the aesthetic or spiritual value of forests. Trees may still be logged but only on a careful, selective, sustained-yield basis compatible with the health of the whole system over the long term.

A second example, Balinese agriculture, illustrates some of the dangers of intervening in well-established systems. It also suggests that such systems possess a kind of wisdom—or as Bateson would have it, sanity—that we are just now beginning to rediscover and appreciate. Rice growing on Bali was regulated for centuries by what is tantamount to a hydraulic religion. The result was a stable and efficient agricultural system that for over a millennium produced reliable crops while preserving the fertility of the soil.

Development-minded outsiders, however, scorned the old ways as superstitious and persuaded or bribed peasants to convert to the so-called green revolution—the new religion of mechanical agriculture, which mandated intensive and continuous cropping of special high-yielding strains of rice supported by heavy applications of fertilizer and pesticide.

Although yields surged at first, they soon began to decline to previous levels as pests developed resistance to the chemicals used to suppress them. In addition, there were many costly "side effects." For example, due to the toxicity of the chemicals, eels and fish could no longer be raised in the paddies, so peasants lost an important traditional source of protein. The green revolution may have succeeded elsewhere, but it failed on Bali, and the Balinese soon reverted to the ways of their ancestors.

Intrigued by this unexpected outcome, an anthropologist and a biologist constructed a computer model of the traditional agricultural regime. This demonstrated conclusively that the centuries-old method of growing rice was, in fact, very close to the ecological optimum.[27]

The Balinese were lucky. Their traditional system was not destroyed or forgotten during their brief flirtation with the new religion, and reconversion was therefore easy. Others have not been so fortunate. Perhaps the most tragic instance of well-meaning intervention gone awry comes from the arid African Sahel, where mechanically minded experts, convinced that progress was being impeded by the "obvious" lack of water, drilled artesian wells for the local pastoralists. But the expansion of flocks and herds made possible by abundant water proved to be a fatal improvement. More animals than the whole system could support year in and year out led to catastrophe—ruinous overgrazing, social chaos due to the destruction of the traditional system for allocating water and pasturage, and an eventual famine of major proportions. A seemingly benign intervention destroyed a culture that had sustained the Sahelian peoples for generations—meagerly but with dignity.[28]

Beyond the superficial morals to be drawn from this sad tale—good intentions are never enough, and doing the obvious is not always the best course of action—lies a deeper lesson: limits have a function. In this case, the limiting factor identified by development experts as the problem to be solved was, in fact, the linchpin of the society. Since the traditional society was organized around water scarcity, flooding it with water was bound to destroy it. To reiterate a point made in previous chapters, when it comes to systems, limits may not be barriers to overcome but boundaries to be respected.

One final example shows how systems may not only defeat unsophisticated analysis and intervention but can also eventually repulse all efforts to bend them to human will. Indeed, there seem to be certain evils that we must simply endure because they are an intrinsic part of the system. To keep struggling against them wastes time, money, energy, and credibility. A classic case in point is the tsetse fly. Efforts to eradicate this so-called pest and thereby open all of Africa to human settlement after World War II failed. It was finally realized that nothing short of exterminating every antelope, gazelle, and zebra could effect its eradication, so the areas infested by tsetse were grudgingly left to wildlife. Ironically, the magnificent fauna of Africa may owe its survival in good part to this pestiferous insect. Again, what mechanical mind identifies and attacks as a problem will appear quite different to ecological mind, which takes it as axiomatic that all elements of a system are there for a reason and that the so-called side effects can never be separated from the intended effects. Indeed, from the systems point of view, every simple-minded, linear intervention in the web of life is likely to produce detrimental consequences, now or in the future.

Much more could be said about the principles of system behavior and about the many traps and paradoxes that systems generate and about the many ways shortsighted, neurotic human beings are readily ensnared by them (one reason that personal and societal addiction has become a pervasive feature of modern life[29]). However, the essential teaching of the systems paradigm, says biologist and systems analyst Donella Meadows, is simple:

> The world is a complex, interconnected, finite, ecological-social-psychological-economic system. We treat it as if it were not, as if it were divisible, separable, simple, and infinite. Our persistent, intractable, global problems arise directly from this mismatch.[30]

Many of our most fundamental dilemmas therefore arise not so much because human beings are egotistical and greedy—although these and other moral failings certainly do not help matters—but as a consequence of simple ignorance. Meadows continues:

> No one wants or works to generate hunger, poverty, pollution, or the elimination of species. Very few people favor arms races or terrorism or alcoholism or inflation. Yet those results are consistently produced by the system-as-a-whole, despite many policies and much effort directed against them.[31]

This perverse result occurs because the vast majority of the population—including almost all so-called authorities—fails to think systematically. Witness the way in which we typically chop up the world intellectually into "disciplines" that produce "experts" who know only their own narrow specialty. The will to solve a particular problem may be there, but an understanding of how systems actually behave, as opposed to how we would like them to behave, is lacking. Not only do problems not get solved, but most of the efforts directed at solving them are entirely wasted. In short, ignorance of systems behavior produces a kind of insanity—doing the same thing over and over again even though it does not work.

A classic example is the plight of farmers in a market economy. A farmer succeeds by producing more. But the effect of more production is to lower prices. This drives marginal producers out of business and, paradoxically, requires successful farmers to work even harder to maintain their income. Farmers are therefore caught in a classic systems trap—a vicious circle in which they have to "Get big, or get out." Enter the politicians, eager to solve the problem. But instead of doing the one thing that systems analysis says would actually preserve small-hold farming—the "unthinkable" act of limiting the size of farms—they provide subsidies whose real-world consequence is to enable the successful to get bigger yet. This causes more family farms to slide into oblivion and generates a renewed demand for subsidies, and so on, until only giant industrial farms are left. So our system of agricultural subsidies guarantees the triumph of agribusiness, precisely the outcome we supposedly wanted to prevent.[32]

As this example reveals, ignorance is compounded by interest. Those in positions of wealth and power have a stake in maintaining the current mechanical way of thinking and doing, so they are not going to rock the boat by proposing the "unthinkable." Politicians have no incentive to deal with problems realistically. After all, the painful truth about systems behavior is that there are rarely easy or attractive solutions to problems. As the farming case shows, the most effective solution may be "counterintuitive"—systems jargon for "surprising" or even "shocking"—so no sane politician would vote for it. (Another example is deliberately maintaining high gasoline prices, despite the short-term pain this inflicts, to spur conservation and innovation, which are necessary to break the addiction to oil that caused the high prices in the first place.) Even if politicians understand the ways in which the system itself generates the problem, they are loath to propose real changes that will exact real sacrifices; it is much easier and safer to practice symbolic politics. (The collective refusal of the American political class to confront our addiction to oil and to profligate energy use is a lamentable case in point.) Anyone trying to be realistic and honest confronts the Platonic predicament: few will relish being told that they are fundamentally deceived or that they themselves might be ultimately responsible for the ills that beset them.

To mention Plato is to raise the political challenge of systems. Given everything that has been said hitherto about the nature of complex systems and about the cognitive and emotional limits of passionate human beings, what is the likelihood that most people will "see the light"? Will they dispel their ignorance of systems behavior and enthusiastically embrace the ecological worldview, including the ethical mandates of humility, moderation, and connection that follow inescapably from that worldview?

Unfortunately, the question practically answers itself. As we have previously seen, only a minority seems capable of rising very far above Piaget's concrete operational stage to the formal operational stage characterized by the logical use of symbols in abstract thought. More generally, much as the democratic ethos would wish otherwise, people vary markedly in ability. Although many can cook, few can do theoretical physics. Charles Murray points out the obvious: since by definition all children are not above average, educational and social policies that pretend they are, are literally fantastic and doomed to failure.[33] The eternal political challenge is therefore to ensure that the inevitable elites are prevented from exploiting or oppressing the populace (an issue explored in later chapters).

This political challenge is all the greater today in that it now appears there may be a yet higher stage of cognitive development. The systems operational stage is characterized by the capacity to comprehend the dynamics of complex systems. This capacity is not merely intellectual. As Bateson states and as many others imply, it demands a kind of wisdom——a virtue not always found among those who have mastered formal operations. Our supposed best and brightest are too often one-dimensional exponents of mechanical mind and linear thought.

The question is whether we confront a brute Platonic reality in the form of a brass majority that is stuck at the concrete level, a silver minority that can think abstractly but not systematically, and a gold elite that is not merely intelligent but also wise? Or might it be possible to turn silver into gold by a transformed educational system that inculcates the new *paideia*, including a thorough grounding in the systems paradigm? Whatever the answer to the latter question may be, it seems doubtful that brass will become precious metal anytime soon. We are therefore impelled toward the Platonic conclusion—"philosophy" for the few, a "noble lie" for the many.

This conclusion is reinforced by the findings of depth psychology. According to Jung, a tripartite division of humanity also exists psychologically. A lower stratum exists in a "state of unconsciousness little different from that of primitives," a middle stratum has "a level of consciousness that corresponds to the beginnings of human culture," and a higher stratum possesses a consciousness that "reflects the life of the past few centuries" (and is therefore "modern" in the positive sense of the word).[34] Recall also Jung's conclusion in the previous chapter: the majority, being still essentially "in the state of childhood . . . needs authority, guidance, law" as well as the "crutch" of a collective identity animated by a guiding mythos—in other words, a noble lie.

Before addressing Plato's shocking juxtaposition of nobility and falsehood and its possible meaning for our times, I should first discuss my approach to *The Republic*. Any statement about the so-called fountainhead of Western philosophy will inevitably be controversial because *The Republic* has many levels of meaning and also presents formidable barriers to understanding. Nothing I say here will end the controversy. However, my aim is not to assert the rightness of my interpretation—a full defense would require another, different book—but instead to use Plato as a vehicle for exposing our current epistemological, ethical, and political plight.

The treachery of translation from ancient Greek into modern English is only the first of many daunting problems in approaching *The Republic*. A more serious obstacle is that it is difficult, if not impossible, for us to project ourselves back into a cultural mindset and historical setting very different from our own. To cite just two major differences, ancient Greece was still primarily an oral culture making a troubled transition to literacy, while we are still predominantly a literate culture rapidly moving toward visuality; and we are hyper-rational, whereas the ancient Greeks, for all their newfound intoxication with reason, were still quasi-shamanic.[35] Moreover, Plato is essentially a mystic seeking enlightenment, not a worldly philosopher pursuing secular knowledge. Another of his dialogues shows that he has more in common with Gautama Buddha than Karl Marx, so he must be read accordingly.[36] In short, the cultural and intellectual distance between Plato's *Republic* and us is enormous.

Perhaps the most formidable barrier to understanding *The Republic* is the intrinsic nature of the dialogue form. A Platonic dialogue is not a straightforward argument of the kind we are accustomed to but a philosophical drama that demands an attentive and astute reader who can follow the twists and turns of the conversation to penetrate to its core meaning, which may be implicit rather than explicit. To make matters worse, Plato's Socrates, the protagonist of the drama, is not only a supreme ironist—nothing he says can be taken at face value— but also a trickster who delights in pulling the philosophical rug out from under his interlocutors, building up position after position only to tear them down later.

In other words, *The Republic* is not a political program.[37] The modern tendency to interpret it anachronistically and ideologically—as if it were *The Communist Manifesto* or a utopian tract advocating selective breeding, enforced poverty, and the dictatorship of a philosopher-king—is completely mistaken. The elaboration of the ideal state by Socrates is not to be taken at face value. A careful reading shows that Socrates creates the city of perfect justice *as a thought experiment* to demonstrate the impossibility of constructing a truly just regime or of preserving such a chimera from rapid decay if it were to be built.

Plato's great dialogue is a work of philosophy, not ideology. Those who see in it a justification for tyranny have misunderstood Plato.[38] Socrates's aim is to teach political moderation and philosophical dispassion to his young interlocutors—that is, to incline their minds to wisdom and virtue instead of ambition for wealth, honor, and power. Far from being a utopia, except in the sense of providing a standpoint from which to judge lesser regimes, *The Republic* is, as Allan Bloom says, "the greatest critique of political idealism ever written . . . serv[ing] to moderate the extreme passion for political justice by showing the limits of what can be demanded and expected of the city."[39]

To speak more generally about the premodern approach to politics, of which Plato's Socrates is the shining exemplar, the thinkers of antiquity regarded the pursuit of perfection in the political, social, and human spheres as simply mad—as hubris—because humankind can no more achieve universal justice and perpetual peace than it can abolish droughts and hurricanes. The "political animal" is a mortal, not a god, and therefore is imperfect by definition. This does not mean that premodern thinkers lacked ideals. They usually had high aspirations for humanity—indeed, far higher than our own—but for the most part they avoided ideal*ism*. That is, they refrained from externalizing the drive for perfection into a political ideology and instead turned the drive inward, aiming primarily at *self*-perfection in an inherently limited, imperfect world.

With this preamble, we can return to the noble lie—or as it is sometimes translated, "noble fiction." Plato's argument starts from the premise that we are in a kind of epistemological black hole from which escape is almost impossible (a premise that has been confirmed by modern science). With our normal waking consciousness, we perceive only the shadows cast on the screen of the mind by the senses. Since the shadows by themselves make no sense, we must give them order and meaning by making up stories about them—what Jeans called "parables" and Socrates, with more candor, called "fictions" or "lies."

Some of these stories are better than others. They come closer to what Plato called "true opinion," meaning that they give a reasonably accurate account of a particular region of shadows. This is the aim and function of science—to provide stories about physical phenomena that are both accurate and useful. As we have seen, however, even scientific research is not purely objective. It is not the facts that evince the truth, but the metaphors, paradigms, and theories that decide the facts—so even scientific accounts are still only parables, not absolute truth. [40]

Because true opinion is not a reasonable or achievable goal in most areas of life, our values and purposes, not some elusive canon of "truth," determine how we order the shadows. The choice is ultimately moral and aesthetic, for the story we tell will entail a way of life that is more or less beautiful, harmonious, and just. The pith of Socrates's teaching in *The Republic* is that whatever we say about the shadows will be a lie, so we have only a choice among lies. Because our lives will be beautiful or ugly according to which lie we choose, we need to choose a noble lie that uplifts us and encourages us to pursue wisdom, virtue, and justice rather than self-aggrandizement or material gain.

In effect, living in a cave of shadows condemns us to be poets on a grand scale—poets of life, whose canvas is the human community. Socrates's famous quarrel with the poets in *The Republic* is usually misunderstood as expressing Plato's aversion to poetry itself. In fact, his condemnation is directed at bardic culture—that is, the particular themes and metaphors employed by Homer and other popular poets—because he understood how powerfully it shaped the minds of his compatriots in perverse ways. Although it was not Plato who said, "Poets are the unacknowledged legislators of the world," Shelley's aphorism would seem to express Plato's sentiments exactly.[41] From the metaphor of the pilot in the beginning of his dialogue, to the image of the cave in the middle, to the myth of Er at the end, Plato is a poet engaged, like any other poet, in a search for fresh metaphors that bring new light into the cave—and therefore the possibility of a better and more conscious life, both for individuals and for the polis. Platonic politics is thus a special kind of poetry.

The powerful image of the noble lie addresses a deep political conundrum. To cast the conundrum its Platonic form, philosophy is politically dangerous because it can lead one to doubt the common opinion—that is, the validity of the religious and social beliefs or even the laws of one's community. If practiced by the ordinary person, philosophy will therefore cause the erosion of the mythos and ethos on which the polity rests. (As the fate of Socrates himself illustrates, the community is likely to regard philosophy as a criminal pursuit because it calls into question ruling norms and conventions.) And philosophy is like pregnancy in being an all-or-nothing affair. One cannot combine a little bit of philosophical knowledge with a lot of popular opinion, for one simply gets the worst of both worlds—neither wisdom nor stability.

The average person will never be a rough-and-ready philosopher capable of living autonomously, authentically, and virtuously in a mythless world. The "2,000,000-year-old man" cannot be satisfied with a merely scientific account of reality. He must have poetry, whether good or bad, because poetry provides *meaning*, whereas science gives only *knowledge*. So Socrates shifts ground in mid-dialogue. He abandons the ambitious project of constructing the just regime by reason and instead settles for the much more modest goal of replacing foolish and defective popular lies with nobler ones—that is, with political beliefs that are likelier to foster peace, virtue, obedience, and a rough kind of justice than the inherited social fictions. The conclusion of *The Republic* is a story that Socrates says, "could save us, if we were persuaded by it." The myth of Er epitomizes in dramatic form the lesson of the work as a whole—that the purpose of life, and therefore the vocation of politics, is soul making.[42]

The Republic thus ends on a note of irony. Socrates's self-styled "lie" is what he "knows" to be "true" (through *gnosis*, the direct apperception of the reality outside the cave). That this truth cannot be verified scientifically is beside the point, for to the extent that it encourages men and women to choose a nobler way of life—a life marked by beauty, grace, harmony, proportion, and other virtues—then it is morally "true."

Whatever one might think about Plato's argument, the conundrum itself is unavoidable and has haunted Western thought ever since. It is, for instance, at the core of Jean-Jacques Rousseau's oeuvre, clearly but paradoxically articulated in his famous assertion that man must be "forced to be free" by a rigorous political education and a willing obedience to civic religion. It is also found in the story of the Grand Inquisitor from Dostoevsky's *The Brothers Karamazov*. The Inquisitor admonishes Christ for having mistakenly given people spiritual freedom when what they really need is "miracle, mystery, and authority."

We have in effect replicated Plato's thought experiment in real life during the past three centuries. The myths that formerly sustained Western civilization have been almost entirely dissolved in the acid of instrumental rationality, thereby corroding away the moral basis of the polity. To use Socratic language, the lie that underlies modern life has proven to be ignoble—to be a real lie instead of a useful and necessary social fiction. It claims that human beings are capable of rationally understanding and controlling both organic and human nature and of using the powerful means that rationality provides for mostly benign or even utopian ends.

But history has belied this claim. The drive to dominate nature has generated a vicious circle of ecological self-destruction, an excess of rationality has unleashed a host of irrational forces that are the shadow side of that same rationality, the pursuit of material gratification has spawned addiction and frustration, and the struggle to control economic complexity and social demoralization has fomented an ever growing concentration of political and administrative power. Thus "progress" has proven to be a myth in the pejorative sense of the word.

The political challenge of our time is almost exactly that mooted by Socrates in *The Republic*. The attempt to live "scientifically"—that is, to rely on reason unalloyed by myth—has failed. The political animal simply cannot exist without some kind of story that gives meaning and coherence to life and that provides the intellectual and moral basis for political community. This would be true even if we could raise the whole human race to an unprecedented level of intellectual, psychological, and ethical maturity. We have no alternative but to go in search of a nobler lie on which to reconstitute the polity.

We immediately encounter the paradox that myth, by its very nature, cannot be created rationally. It necessarily bubbles up from below, from the depths of the human unconscious. So it must emerge organically. The task is not to invent a myth and then impose it on the masses—an impossibility—but rather to end the tyranny of "overvalued reason" by restoring mythopoetic thought to intellectual respectability.

I do not use the word *tyranny* lightly. When a purely instrumental rationality—reason unaided by art, myth, and the like—dominates, the outcome is bound to be a society that is radically unequal. Because only a minority is capable of formal operations, much less systems operations, roughly two-thirds of the population is automatically excluded from the meritocratic race for wealth and status. So the egalitarianism of modern civilization is largely a sham. Except at the juridical level, and not always even there, the game is rigged against the average man or woman. However, if instrumental reason were complemented by the reasons of the heart in general and mythopoetic consciousness in particular, the culture would be more balanced—psychically, socially, and economically. When it comes to the mythic realm, we are all on a relatively equal footing. Along with concrete operations, myth is what the average human psyche was constructed to understand. In short, although we cannot survive without a cognitive elite, the restoration of myth would help to level the cultural playing field for the less academically talented.

A detailed discussion of the educational reforms needed to achieve this end or to institute *paideia* in general is beyond my scope. However, Howard Gardner's *Frames of Mind* is a good place to start. We have multiple "intelligences"—verbal, logical, musical, spatial, bodily, personal—and must educate for all of them, not just for reason narrowly defined. Gardner suggests various ways to lift education out of its academic rut—for example, reviving apprenticeships, reinstituting initiations, and perhaps even restoring the "bush schools" of our archaic ancestors. Again, Pascal's heart is an organ of apperception that responds to the totality of life and must be trained in all its aspects or it will atrophy.[43]

If we cannot invent a new myth with our rational mind, that does not mean that we should simply let nature take its course and do nothing to encourage or shape its emergence. If we do not do so, William Butler Yeats's "The Second Coming" may prove prophetic. If the birth of the new myth is resisted instead of assisted, then the likely consequence of today's "mere anarchy" will be a "rough beast"—that is, a violent upsurge of unconscious forces producing an irrational and messianic myth that oppresses rather than liberates. We are called to be the midwives of a benign myth that expresses and supports the new *paideia*.

The argument thus far has been a systematic attempt to show that a new morality and a new myth are both necessary and natural. Contrary to the claims of the secular rationalist, universal moral principles are discernible by reason even though they cannot be proved by logic. Humility, moderation, and connection are enjoined on us by the very nature of the universe we inhabit.

These and other moral or spiritual principles are actually common knowledge and shared by almost all cultures even though they are not always commonly observed. For instance, despite being rooted in Hinduism and Indian culture, Gandhi's political vision was in essence Socratic. His five great principles—*swaraj, satya, ahimsa, satyagraha,* and *sarvodaya*—were taught and exemplified by Socrates more than two millennia earlier. Like Gandhi, Socrates urged self-knowledge and self-mastery, was in a constant search for the truth, preferred to suffer evil rather than do it, stuck fast to his truth even though it cost him his life, and lived a life of simplicity and service. In other words, the vision of *paideia* sketched above is not exceptional or parochial. On the contrary, it would seem to respond to the deepest moral sentiments of humanity, whenever and wherever found.

With respect to myth, we have in the course of the argument accumulated a number of elements of thought that can serve as mythologems—that is, motifs around which a new myth could crystallize. For instance, ecology as master science and metaphor leads quite naturally toward an acceptance of Gaia as a mythological symbol of scientific truth. Or the universality and centrality of consciousness in the life process effectively reenchants the world and restores the I-thou relationship between man and nature asserted by ancient myths. Or individuation as a path of personal and social evolution toward a deeper consciousness and a more balanced wholeness restates the universal mythologem of "the hero's journey."

What is noteworthy about these and all the other elements discussed above is that they are not fantastic. At least in the first two cases, they fall squarely in the category of "true opinion"—that is, demonstrable knowledge (in the form of "parables," to be sure) concerning the shadows on the cave wall. Nor are they tribal—like the myths of antiquity, which too often exalted one particular folkway or religion over all others. So we are not forced to believe in absurdities in order to possess a living myth. The story of Gaia may seem spooky to the hardcore rationalist, but it is not a fairy tale, and it conveys an essential truth that we have forgotten and must now relearn.

To conclude, we have all the makings of a powerful, beautiful, and authentic story of who we humans are and what we should do with our lives—a story that would foster cognitive and intellectual sanity by reconnecting us to the aesthetic unity of the cosmos. There remains the fraught task of translating this vision of the new *paideia* into a workable and humane politics: how shall we constitute a politics of consciousness grounded in ecology?

6
Politeia

[An individual] might wish to enjoy the rights of a citizen without wanting to fulfill the duties of a subject, an injustice whose spread would cause the ruin of the body politic.

Therefore, in order for the social compact not to be an ineffectual formula . . . whoever refuses to obey the general will shall be constrained to do so by the entire body; which means only that he will be forced to be free. . . .

For the impulse of appetite alone is slavery, and obedience to the law one has prescribed for oneself is freedom.
—Jean-Jacques Rousseau[1]

Politeia is the means for realizing the ends of *therapeia* and *paideia*. Wisdom and virtue do not arise spontaneously in human beings, especially those who reside in complex civilizations, so morality must be institutionalized and inculcated by a polity dedicated to fostering and upholding society's norms and mores. The polity's role is to govern—to direct affairs in a way that citizens are encouraged to follow a moral code or are swiftly checked when they fail to do so. (All the rest of what we call politics is politicking, policing, and administration.) Provided that the code reflects an elevated ideal, such as excellence or wisdom, the result will be a rule of life that is relatively sane and humane.

We live at a historical juncture that will challenge governments as never before. Liberal society owes its existence to the bubble of ecological affluence fueled by the "discovery" of the New World and the exploitation of the stored solar energy in fossil fuels.[2] Those who have grown up in affluent societies have therefore enjoyed unprecedented opportunities and freedoms, as well as levels of comfort and plenty all but unimaginable to our ancestors. But the impending return of ecological scarcity means that the expectations and aspirations of billions of individuals cannot be met and that individual wants will increasingly be subordinated to collective needs. Governments now confront the Herculean task of effecting an epochal economic, social, and political transition from the industrial age to the age of ecology. The question is whether this can be achieved without lapsing into totalitarian tyranny or religious despotism.

To escape such a fate will require us to break decisively with our habitual response to societal problems: passing laws that give governments ever more administrative power. The extraordinary nature of the challenge exposes us to the eternal and inescapable dilemma of politics in a particularly acute form. As Lord Acton observed, because power inevitably corrupts, no can be entrusted with it: "The danger is not that a particular class is unfit to govern. Every class is unfit to govern."[3]

The maxim "That government is best which governs least" follows as a matter of course. As John Stuart Mill says in concluding *On Liberty*,

> A government cannot have too much of the kind of activity which does not impede, but aids and stimulates, individual exertion and development. The mischief begins when, instead of calling forth the

activities and powers of individuals and bodies, it substitutes its own activity for theirs.[4]

So the proper function of government is to facilitate, not dominate; to make the rules, not play the game. By its nature, big government—whoever exercises power and whatever their intentions—is bound to be less responsive or efficient than small. In addition, any problems that emerge quickly become the rationale for further extensions of administrative power. But the more the government intrudes into the life of the citizenry, the more burdensome and expensive it becomes. More important, because power corrupts, it will inevitably tend to become overbearing as well.

Because men and women have a surfeit of passion and a deficit of reason, a substantial degree of governance is indispensable for civilized life. It alone can constrain the one and supply the other. So government is a necessary evil—and the more it departs from what is truly indispensable, the greater the evil. Our aim must therefore be to construct a political regime that is sufficient to the desired end without exceeding what is strictly necessary.

Instituting a necessary evil is not for the squeamish, but we shirk the task at our peril. Pascal likened political philosophy to "lay[ing] down rules for an insane asylum."[5] The metaphor is apt. At best, even in comparatively well-ordered polities, political life is a kind of Bedlam characterized by shared delusion, cold-blooded self-seeking, and an aggressive will to power. At worst, it becomes a barely sublimated civil war one step removed from a Hobbesian state of nature. And Pope to the contrary notwithstanding, good rules are indeed necessary for a good politics lest we turn into a well-administered concentration camp. Depending on the rules and the ways in which the rules are administered, the asylum will be more or less peaceful, more or less benign, and more or less conducive to individual sanity and welfare.

Politics is not everywhere and always an unmitigated evil. As Aristotle and others point out, participation in politics can enhance the self-development of individuals. What is too often forgotten, however, is that only small, simple, face-to-face societies permit genuine participation. In the wrong setting—a society that is large, complex, or divided—participatory politics is likely to become what Plato said it was: an ignorant and impassioned mob fighting over the tiller of the ship of state, with potentially disastrous consequences. The essential task is therefore to foster a social and economic setting conducive to a politics that is sane, humane, participatory, and ecological.

Nothing I say here should be construed as approving a dictatorial remaking of our civilization. We do not need a Lenin or even an Ataturk. We require a new moral, legal, and political *order* that cannot be imposed from the top down but that must instead percolate up as the consequence of an intellectual and moral reformation. The aim of this reformation should be to create the kind of society desired by Burke and Taine—a self-regulating society in which individuals bind themselves with moral chains and thereby become their own constables.

To return to the theme of the noble lie, the ideal animating the machinery of government, not the machinery itself, constitutes a polity. Institutions do not create an ethos: witness the bootless attempts in the postcolonial era to graft the trappings of representative democracy onto traditional societies for whom democracy and liberty are alien ideals. The reverse is actually true: those who possess an ethos will naturally establish institutions that reflect it.

Politics is not about elections, offices, or laws. It is about the definition of reality: what epistemology, ontology, and ethic shall constitute our rule of life? It is about the master metaphor that frames the manner of thought and the character of institutions at lower levels. At the heart of any political battle—from the general direction of society to particular policy issues—is a fight to make a particular idea prevail: the invisible hand or the class struggle, a right to life or freedom of choice?[6]

In consequence, said David Hume, it is always opinion that governs:

NOTHING appears more surprising to those who consider human affairs with a philosophical eye, than the easiness with which the many are governed by the few. . . . When we inquire by what means this wonder is effected, we shall find, that, as FORCE is always on the side of the governed, the governors have nothing to support them but opinion. It is, therefore, on opinion only that government is founded; and this maxim extends to the most despotic and most military governments, as well as to the most free and most popular.[7]

The French jurist J. M. A. Servan made the same point more cynically:

A stupid despot may constrain his slaves with iron chains; but a true politician binds them even more strongly by the chain of their own ideas. . . . [T]his link is all the stronger in that we do not know what it is made of.[8]

This ancient problem was adumbrated by Plato in *The Republic* and has been much studied in modern times by sociologists of knowledge: the human mind produces opinions that may have only a passing resemblance to reality. In the political arena, the problem manifests as Karl Marx's "false consciousness." On one side, a majority is unskilled in thinking but hungry for meaning, and on the other, a smaller minority is skilled at mental manipulation and hungry for power. The latter normally succeeds in imposing its ideas on the former—"What luck for rulers that men do not think," said Adolf Hitler[9]—and these ideas, backed up as necessary by the gendarmerie and secret police, constitute the mainstay of any regime.

The political dramas that occupy our newspapers and television screens are therefore largely irrelevant. As long as the basic metaphor remains the same, it is business as usual, no matter who wins elections or what policies are adopted. However, let one metaphor displace another, and "reality" shifts accordingly. According to Archibald MacLeish, "A world ends when its metaphor has died."[10] When the consent of the governed supplanted the divine right of kings as master metaphor, the consequence was a radically new and different political order.

The usual understanding of false consciousness, especially among Marxists, is that the falsity is due to cynical political manipulation combined with deliberate intellectual obfuscation, which leads the masses to be cunningly imprisoned in a set of beliefs that serve the interests of the ruling class. The usual solution proposed by modern thinkers is therefore scientific. Science—whether the laws of dialectical materialism that govern the unfolding of human history, the best means of fostering economic growth, or the right way to feed babies—will provide objectively truthful answers to all questions and thereby liberate us from false consciousness once and for all.

But there is no such thing as objectively true consciousness. Science may indeed provide us with true opinion concerning certain aspects of human nature and the natural world so that we can choose a rule of life that does not flout reality. But it cannot tell us what reality ultimately is, and it cannot choose the rule for us. Everything depends on the master metaphor we use to construct reality. The image of the machine leads to one kind of society—individualistic, acquisitive, exploitative—whereas the image of Gaia points in a very different direction. Again, we come face to face with the enormous power and reach of metaphor. It can liberate us, or it can enclose us in a mental prison—either one of our own making or one imposed on us by powerful others.

The essential political struggle of our time is not to pass laws that reduce pollution and conserve energy so that the machine can keep running until it self-destructs, taking humanity along with it. Instead, it is to fight to make ecology the master science and Gaia the ruling metaphor—to abandon an ignoble lie and embrace a nobler new fiction that offers the means of long-term survival and the prospect of a further advance in civilization.

This conclusion that a new fiction is the key to political change is supported by systems analysis. As Donella H. Meadows points out, the most effective leverage point for changing a system's behavior is its fundamental mind set or paradigm, for this determines its goals, structure, and rules.[11] Unfortunately, this is also where resistance to change is fiercest. The required strategy of change, says Meadows, is to expose the anomalies, contradictions, and failures of the old paradigm while at the same time offering a new and better one.[12]

The essence of the *politeia* that follows from this new fiction has already been stated. It is a politics of consciousness grounded in ecology and dedicated to inner cultivation instead of outer conquest. But what does this imply?

The sages, prophets, poets, and philosophers who have gone against the grain of civilization by urging men and women to pursue wisdom and virtue instead of wealth and power have generally agreed on the means necessary to this end. They all envisioned a way of living that is materially and institutionally simple but culturally and spiritually rich—and therefore more generally free, egalitarian, and fraternal than life in complex societies devoted to continuous accumulation and expansion.

The case for material and institutional simplicity takes several forms. The negative argument is that as societies grow larger and more complex, self-regulation breaks down, so they develop chronic and intractable problems. Politicians respond with laws and regulations that purport to be solutions. But when society has reached a certain level of complexity, solutions are either far from obvious or too painful to implement or even contemplate. Leaders resort to simplistic, merely expedient "reforms" that fail to solve the old problems and generate new ones that then require stronger measures. As a consequence, government grows first powerful, then intrusive, and finally overbearing or even tyrannical, and the people themselves are corrupted and made dependent. Under such conditions, liberty decays, equality declines, and fraternity fades, often dramatically. The solution is for men and women to live in relatively small and simple societies that encourage them to be upright and independent, that preserve them from oppression, that keep them on a relatively equal footing with their fellow citizens, and that allow them to participate meaningfully in civic life.

The positive argument is that men and women should live close to the earth and to each other in relatively simple and stable small communities because this is what the archetypal needs of the "2,000,000-year-old man" require. A simpler and more natural existence will tend to maximize an individual's chances of enjoying the good life—defined as a way of living that is filled with nature, beauty, family, friendship, leisure, education, and, for those inclined to it, philosophy in the Platonic sense of personal and spiritual self-development. These things, not material goods, bring true felicity.

It follows that the aim of economic life must be sufficiency, which supports such felicity—not great wealth, which is its enemy. Sufficiency is also important for political reasons. Besides forestalling the growth of tyranny, a simple economy is relatively transparent, so individuals can see their own interests as well as the common interest and act on them. Sufficiency combined with ample opportunity for self-development also reconciles the tension between equality and excellence. If each human being attains his or her unique excellence and is recognized by others for having done so, then the best can in principle rule without creating either dependency or resentment among the ruled.

This brief overview touches on important issues that are further addressed below, but we must first respond to the objection that a small-is-beautiful prescription for political salvation is utterly utopian and therefore not worthy of being taken seriously. In fact, what has always been philosophically commendable is about to become practically obligatory. The manifold pressures of ecological scarcity will soon compel us to live in smaller, simpler communities that are closer to the land than the megacities of industrial civilization. In the next few decades, well before we have completely exhausted the capital stocks of fossil fuels and mineral ores on which the current industrial order depends, matter and energy will become increasingly scarce and expensive. If deployed skillfully and in a timely manner, technology can shape and moderate this inexorable trend, but it cannot forestall it. Our future way of life will of necessity be more simple, frugal, local, agricultural, diversified, and decentralized than at present. Our task must be to make a virtue of this necessity.

When we recognize its necessity, we shall see that a simpler way of life might indeed be more virtuous and happy than the one we now believe represents the acme of human progress. In the first place, industrial civilization has become too complex and interlinked for its own good. As Joseph Tainter points out, an excess of complexity, usually aggravated by other factors, has spelled the downfall of previous civilizations.[13] The costs of increasing complexity grow disproportionately until they eventually reach a point of diminishing or even declining returns. The civilization therefore has to run harder and harder to make further progress or even to stay in the same place.

In addition, a civilization already stressed by the high costs of complexity may no longer be resilient enough to respond to further challenges. It risks a cascade of failure should a critical link fail for whatever reason. The interconnected institutions of a highly complex society are like mountain climbers tied to one rope with no belay: the fall of one can trigger the death of all. For example, the world financial system experiences periodic crises when the failure of one bank brings down a host of counterparties. Similarly, a sudden or significant increase in the price of a critical commodity, such as petroleum, can choke an industrial superstructure predicated on cheap and abundant energy. The further danger is that such a crisis can trigger psychological panic and social pandemonium. In short, the higher we build the edifice of civilization, the more vulnerable we become to catastrophe. A simpler, more resilient way of life would therefore be advisable on prudential grounds alone.

But our primary concern here is *politeia*, and the political argument for cultural simplicity is that great size and complexity produce a debased politics. When a polity grows beyond certain bounds, oligarchy in the bad sense is inescapable, the burden of bureaucracy grows ever more stifling, and genuine consent of the governed is practically unattainable. A vicious circle fostering ever greater centralized planning, administrative intervention, and political control takes over. If democracy survives at all, it will be a token democracy shadowed by the lurking menace of mob rule.

In the United States today, for instance, a tiny circulating policy elite makes all the important decisions in ways that align the interests of government, finance, and business. Since the system is "democratic," the elite has to take into account the passions of the mob, which can erupt if its ox is palpably gored. So as long as the American ruling class provides the bread of affluence and an entertaining media circus, it can do pretty much as it likes. Having long since outgrown the relatively simple conditions required to support its constitutional design, the United States has therefore become an imperial polity bearing no resemblance whatsoever to the original American republic. Such is the political price of great size and complexity.

To cast the problem in more philosophical terms, let us turn to Jean-Jacques Rousseau's *On the Social Contract*, which argues that the central task of politics is to uphold the "general will." This is what any reasonable person, putting aside his or her prejudice and self-interest, would agree is in the public interest because it benefits the community as a whole. Rousseau contrasts the general will with the "will of all," which is the mere summation of all the private wills of the individuals composing the polity.

The difference between the general will and the will of all is best seen through examples. If people are carrying a contagious disease, the general will may demand that they be quarantined in some fashion. We do not allow a Typhoid Mary to work in restaurants because preventing the spread of illness to the general population trumps her loss of liberty. Similarly, we do not permit individuals to urinate and defecate just anywhere. We oblige them to practice good hygiene by using sanitary facilities, both to prevent a public nuisance and to preserve public health. We also make immunization mandatory for schoolchildren because we know that the gain to society from herd immunity outweighs not only parental preference but even the slight risk of harm to any particular child. In this critical area of public health, we compel individuals to follow the general will rather than their private will because to do otherwise would produce a diseased will of all.

In these cases, the difference between the general will and the will of all is clear, and the argument for the former is, to most people, compelling. However, this same dynamic applies at every level within the polity—albeit usually in a more attenuated form that can make it hard to achieve or even discern the general will, especially in advance. As Rousseau points out, "One always wants what is good for oneself, but one does not always see it."[14] Even where there is no evil intention but simply the natural urge to fulfill individual desire, people following their private will almost always create a will of all contrary to the general will.

For example, the individual preference for private automobiles leads to a host of public ills—traffic-choked and polluted cities that are friendlier to cars than people, thousands of dead and injured people every year, the threat of climate disruption, the loss of good farmland to suburban sprawl, foreign policy dilemmas or even wars, and so forth. Similarly, private demand for exotic woods causes the destruction of tropical rainforests, an ecological tragedy whose costs we all bear. Likewise, individuals seeking longer life through state-of-the-art medical care threaten to bankrupt the public purse, to mention only the fiscal cost of extended life spans. In other words, perfectly reasonable and legitimate private desires and actions aggregate into global outcomes that no reasonable person would want. Unless the general will is identified and upheld by the polity, ill-advised microdecisions motivated by private interest will add up to an unwanted or even ruinous macrodecision.

The "tragedy of the commons," the "public-goods problem," the phenomenon of "market failure," and a number of other dilemmas much studied by contemporary social scientists are instances of the general problem identified by Rousseau. The same essential conflict occurs within each individual human being. We all know we would be healthier if we ate less and exercised more (the general will), but instead we indulge appetite (the private will) and cause an epidemic of obesity (the will of all).

As a matter of both principle and practice, modern political economies are based explicitly on the will of all—that is, they are designed to satisfy private desire, not to achieve the public good. To put it the other way around, the public good has been redefined as the outcome of the invisible hand of the economic and political marketplace. In fact, any attempt to uphold the commonweal is likely to be dismissed out of hand as special pleading or denounced as hostile to liberty. The practical outcome of modern political economy is almost bound to be what economist John Kenneth Galbraith called "private affluence and public squalor"—that is, a state in which individuals gratify their petty desires without regard to the unwanted or even destructive consequences of their private acts.[15]

Worse yet, the reality of any marketplace is that participants constantly strive to tip the invisible hand in their direction, so the legislative process is likely to be subverted. As Rousseau put it, "the basest interest brazenly adopts the sacred name of the public good . . . and iniquitous decrees whose only goal is the private interest are falsely passed under the name of laws."[16] The resulting will of all is therefore not pure but crooked. It has been bent to favor some interests over others.

For Rousseau, the will of all is not primarily a practical problem to be solved but a moral failure to be overcome. When we follow our private will oblivious to or even in defiance of the general will, we injure society and degrade ourselves. His conclusion is expressed in stark terms by the epigraph to this chapter: since "the impulse of appetite alone is slavery," we must be "forced to be free" by being made obedient to laws that align our private wills with the general will.

Rousseau attempts to reconcile the obvious conflict between individual liberty and the higher freedom we gain in following the general will by saying that we are obeying laws that we, as reasonable beings, have prescribed for ourselves. But he acknowledges that the problem is like squaring the circle—ultimately unsolvable. As Rousseau says, it is simply a given of the human condition that "the private will acts incessantly against the general will," so it is inconceivable that the two will ever be perfectly aligned.[17]

But there is an approximate solution for squaring the political circle: by simplifying the setting of politics, we can make the private will and the general will coincide to a much greater degree than they do in large and complex societies. Rousseau's political ideal is a gathering of peasants deciding their simple affairs under an oak tree. The smaller and simpler the polity, the more likely it is that those deciding will understand the issues, see what would best serve their mutual interest, and choose to implement this collective decision even if it does not fully satisfy their private preferences. There is an almost mathematical relationship: the further away a society is from Rousseau's ideal, the less apparent or compelling the general will is to any given individual, and the greater the likelihood of the polity's lapsing into an undesirable will of all. In short, if you want to achieve a rough approximation of the general will, make your polity small and simple.

It follows that the setting of politics is crucial. Rousseau does not want a totalitarian reign of virtue, as some critics allege. He uses the doctrine of the general will not to justify authoritarianism but to show why it is necessary to establish social conditions that give rise to a natural reign of virtue. Unless the polity is relatively small and simple, the doctrine of the general will can be perverted to legitimate the tyranny of a majority or the dictatorship of a central committee—precisely what happened during and after the French Revolution.

Rousseau's "law one has prescribed for oneself" is not a statute law to be enforced by the authorities but a moral law that embodies the general will. This makes mores the sine qua non of a good politics. Unless the moral law is given concrete form, individuals will tend to go their own way without regard to the general will. Mores, says Rousseau, are the "unshakeable keystone" of politics.[18]

Unfortunately, in large, complex, impersonal societies beyond any person's ken or control, the temptation to ignore or flout the mores of the community becomes overwhelming. Only a relatively small, face-to-face community can exert sufficient moral pressure to make individuals consistently obedient to the mores that force them to be "free." If you want citizens to be upright and law-abiding, make your polity small and simple.

Last but far from least, freedom for Rousseau is not the ability to gratify appetite but the absence of dependence. As he says in *Émile*,

There are two sorts of dependence: dependence on things, which is from nature; dependence on men, which is from society. Dependence on things, since it has no morality, is in no way detrimental to freedom and engenders no vices. Dependence on men, since it is without order [i.e., it is morally degrading], engenders all the vices, and by it, master and slave are mutually corrupted.[19]

In other words, a large, wealthy, complex, hierarchical social order reduces the vast majority to a state of dependence and therefore destroys freedom. So if you want citizens instead of slaves, make your polity small and simple.

Rousseau's doctrines may sound shockingly "illiberal" to the contemporary ear, but consider John Locke's discussion of freedom:

> The *Freedom* then of Man and Liberty of acting according to his own Will, is *grounded on* his having *Reason*, which is able to instruct him in that Law he is to govern himself by, and make him know how far he is left to the freedom of his own will. To turn him loose to an unrestrain'd Liberty, before he has Reason to guide him, is not the allowing him the privilege of his Nature, to be free; but to thrust him out amongst Brutes, and abandon him to a state as wretched, and as much beneath that of a Man, as theirs.[20]

So even the author of the liberal tradition says that freedom is not "an unrestrain'd Liberty." Despite their considerable differences, Locke and Rousseau therefore agree on this fundamental point: man's private will must be made subject to "that Law he is to govern himself by," a law discovered by "Reason."

To frame the issue in terms of Burke's syllogism, if we do not bind ourselves with moral chains, then others will do the job for us (and not necessarily to our advantage). There must be a structure of benign control to teach self-control—in other words, community mores. Locke, who made a strong civil society the linchpin of his political theory, therefore differs only in degree with Plato, Rousseau, and others (such as the psychologist B. F. Skinner) who contend that since social conditioning is already pervasive and controls human behavior unconsciously, we must strive to do it more consciously, compassionately, and responsibly. It also should be said that neither Locke nor Adam Smith would approve of the ends to which their liberal doctrines have been perverted. Both are actually closer in spirit to Rousseau than to contemporary liberals because they envisaged small-hold, independent proprietors enjoying strong but limited property rights, not gargantuan, globe-straddling corporations exploiting the same rights to dominate both economy and polity.

In addition, Rousseau's prescription is very close in spirit and practice to the path not taken in American history—namely, that of Thomas Jefferson, who battled Alexander Hamilton for the soul of the new nation. In essence, they fought to determine which metaphor would govern the American future—Jefferson's garden or Hamilton's machine.[21] As we know, Hamilton won. The mechanical metaphor was enshrined in the U.S. Constitution, and America dedicated itself wholeheartedly to "economic development." Just as the anti-Federalists warned, federal power has steadily eclipsed local authority, commerce and manufacturing have overwhelmed agriculture by industrializing it, and, in general, society has grown ever larger, more centralized, and more complex, causing all of the problems previously enumerated.

Jefferson was not a systematic political philosopher, so he did not argue with Rousseau's rigor, but he reached the same conclusion—the best life for men and women is in relatively small, simple, stable agrarian communities—and for essentially the same reasons. The pith of Jefferson's moral-political argument is found in his *Notes on the State of Virginia*, where he exalted "those who labour in the earth."[22] This was not because farmers are somehow more saintly than burghers but because farmers are *independent*. Those who do not win their subsistence as independent husbandmen, said Jefferson, are condemned to "depend for it on casualties and caprice of customers . . . [which] begets subservience and venality, suffocates the germ of virtue, and prepares fit tools for the designs of ambition."[23] Like Rousseau, Jefferson therefore sees independence as the indispensable basis of liberty.

In addition, because a farmer is almost literally rooted in the local soil, his interest and the community's interest largely coincide. He will therefore tend to be a good citizen supportive of the commonweal, whereas "merchants have no country" and are likely to be both venal and ambitious—that is, they will tend to be bad citizens who seek to profit at the expense of the community.[24] Thus, said Jefferson, "We must make our election between *economy and liberty*, or *profusion and servitude*."[25] It follows that the goal of economics must be sufficiency rather than affluence and stewardship rather than consumption, for the current generation has no right to "eat up the usufruct of the lands for several generations to come."[26]

As a political complement to his modest agrarian economy, Jefferson proposed a system of "ward-republics." Leaving the details of his somewhat artificial and intricate scheme aside, his purpose was to minimize the distance between those who govern and those who are governed to maximize effective political participation and thereby preserve the people's liberty: "Where every man is a sharer in the direction of his ward-republic . . . and feels that he is participator in the government of affairs, not merely at an election one day in the year, but every day . . . he will let the heart be torn out of his body sooner than his power be wrested from him by a Caesar or Bonaparte."[27]

Jefferson, who was well acquainted with Native American life and who styled himself "a savage of the mountains of America,"[28] aimed at creating a civilized analog of the primal band—which he understood to be the original home of liberty, equality, and fraternity—because he believed that genuine democracy could survive only in a setting that fostered a natural sociability. Thus, said Jefferson,

I am convinced that those societies (as the Indians) which live without government enjoy in their general mass an infinitely greater degree of happiness than those who live under European governments. Among the former, public opinion is in the place of law, and restrains morals as powerfully as laws ever did any where. Among the latter, under pretence of governing they have divided their nations into two classes, wolves and sheep.[29]

Rousseau and Jefferson shared a fundamental goal—to enlarge both the quantity and quality of liberty, equality, and fraternity within the polity. And they both agree that the right and proper setting of politics is crucial. Social simplicity and economic frugality are the sine qua non of a relatively benign and democratic politics.

Yet Jefferson was not a democrat as we understand the term. It seemed to him only right, logical, and proper—in a word, natural—that those best fitted to rule should in fact do so. He urged rule by an elite—a "natural aristocracy" of "virtue and talents," which he contrasted unfavorably with an "artificial aristocracy" of "wealth and birth."[30]

This was a sentiment shared by Rousseau, Madison, and even Locke. "In short," said Rousseau, "it is the best and most natural order for the wisest to govern the multitude, as long as it is certain that they govern for its benefit and not for their own."[31] Similarly, said Madison, "The aim of every political constitution is, or ought to be, first to obtain for rulers men who possess most wisdom to discern, and most virtue to pursue, the common good of the society; and in the next place, to take the most effectual precautions for keeping them virtuous whilst they continue to hold their public trust."[32] And Locke's educational proposals were aimed at creating such an elite class, for only those possessed of wisdom and virtue would work to uphold and preserve the civil society that is the foundation of Locke's liberal polity.[33]

Jefferson also regarded education as critically important. Because it was the indispensable means for creating a natural aristocracy, education must be open to all. Otherwise, wealth and birth would become barriers to the rise of virtue and talent. Unfortunately, Jefferson's educational ideal has been largely subverted by the Hamiltonian machine. We educate for merit, not virtue and talent. But a meritocracy is rule by the merely clever, whereas aristocracy is rule by the prudent. To put it another way, a meritocracy adds value—that is, it contributes to the bottom line of the machine without much concern for consequences—whereas an aristocracy adds character, which is essential to keep the train of polity and economy on track. In addition, as a system, meritocracy tends to be self-reinforcing in that the children of the relatively well-off and well-placed will usually enter schools with so many advantages that the eventual outcome is likely to be an artificial aristocracy of wealth and birth.[34]

In the democratic ward-republics envisaged by Jefferson, mere merit in this sense would not easily eclipse virtue and talent. With proper education, or so Jefferson believed, a class of genuine natural aristocrats could arise and be generally accepted by the community. Whatever their level of intelligence or education, people are reasonably shrewd judges of character when they live with each other day in and day out, and as Aristotle said, it is character that counts: "If the citizens of a state are to judge and to distribute offices according to merit, they must know each other's characters; where they do not possess this knowledge, both the election to office and the [administration of justice] will go wrong."[35] In relatively small and local societies, in which everyone knows everyone else's business, citizens will usually agree on who is competent and trustworthy—and elect them accordingly.

In large, complex societies in which men and women know things only at second-hand, however, they cannot judge character and are easily bamboozled by images. Witness the thicket of mendacity and duplicity that is American electoral politics today. Even worse, those who rise to the top of the political heap in such societies will tend to be "ambitious"—that is, avid for power and therefore menaces to liberty.

Although no form of human association can ever be entirely sane and no class is truly fit to govern, a small, face-to-face society in which persons are elevated to office on the basis of direct personal knowledge is more likely to produce political sanity and good governance than one in which consent is remotely engineered by means of images that may have little or nothing to do with reality. In a Jeffersonian setting, natural aristocracy and a democratic politics are allies, not enemies.

To speak more generally, some form of aristocracy is essential for a healthy democracy. As pointed out by Le Bon, only a genuine elite can supply the discipline, prudence, and forethought necessary to check the destructive tendencies of crowds. Without a leavening of aristocratic virtue, any democracy is destined to fulfill John Adams's grim prognosis: "Remember, democracy . . . soon wastes, exhausts, murders itself. There never was a democracy yet that did not commit suicide."[36] And the more massive, crowded, and aggressively egalitarian the democracy, the sooner it will perish.

Although we now understand *politeia* in outline, it would be premature to specify particular institutions. Given the chaotic nature of the world system and the probable magnitude of the impending changes, we confront a future that is both radically different and mostly unforeseeable. Moreover, we are bound to conceive the future anachronistically—that is, in terms of our current way of thinking, which posterity will not share, and our present level of knowledge, which posterity will surpass. Although the future will probably resemble the preindustrial past in many respects, it will also contain possibilities that we cannot now imagine. Technological invention will keep changing the rules of the game, and later generations will almost certainly think and act on premises quite different from our own. Rather than focus on the machinery of government, it will therefore be more fruitful to conceive politics in the light of ecology—that is, to consider how we might model the governance process on the basic physical and biological principles that "govern" natural processes.

This is a complex topic that deserves more extended and careful treatment than I can give it here, but the essence is easily stated. The "constitution" of nature is made up not of structural rules—"Let there be aardvarks!"—but of design criteria, such as the laws of thermodynamics. Nature ordains certain patterns or templates for life—"It shall be based on the chemical properties of carbon"—but it does not mandate particular species. The incredible profusion of life forms constituting the structure of nature is therefore the result of an organic process created by the interaction of all these design criteria over time. We inhabit a Darwinian universe in which the deity did not create a preconceived biological order but instead set in motion a general evolutionary trend, governed by the design criteria, that has churned out species and ecosystems in abundance.

In technical language, nature's strategy is *teleonomic*. Natural systems evolve purposefully toward ends that are implicit in their design criteria, but there is no fixed end toward which the system tends. Nature was quite happy with dinosaurs until an asteroid hit the earth. Yet the outcome is far from random: dinosaurs and mammals both evolved to fill the same ecological niches. By contrast, planning as practiced by humans is normally *teleological* because it involves movement toward a predetermined outcome.

As we have seen, the result of nature's design strategy must be understood dynamically—that is, in terms of positive and negative feedback, oscillation and damping, sensitivity and thresholds, and other identifiable features of systems behavior. Although this is well understood by scientists who study natural processes, it is rare for those concerned with human affairs—social scientists, journalists, politicians, and average citizens—to be even aware that they are dealing with complex systems requiring such sophisticated analysis and rarer still for them to think in a genuinely systematic fashion about polity, economy, and society.

Due to the cognitive limitations previously discussed, we still tend to think about human affairs in ways that are pre-Darwinian. We are preoccupied by mechanics at the expense of dynamics. And we think that change must be imposed from the top down to achieve a predetermined outcome. This way of thinking is obsolete. The natural polis is gone, never to be restored, so we are, as Hobbes said, obliged to create a political community by art. But the mechanical methods of Hobbes and his successors are no longer adequate for the task. To govern effectively today means to comprehend and mimic the organic design strategy of nature.

Architecture, the structural art to which the political art is often compared, provides a concrete illustration of one way in which the problem of political design might be approached. As mentioned in the previous chapter, Christopher Alexander and his colleagues have reconstructed architectural theory on organic principles, arriving at a design philosophy that they call "the timeless way of building," which can be concretely articulated in a "pattern language." One such pattern is "LIGHT ON TWO SIDES OF EVERY ROOM."[37] This pattern does not impose a particular design. An infinite number of rooms—all conforming to the pattern but all different—can be constructed. But it does guarantee a certain result: every room will have enough natural light to feel comfortable. Alexander's group has identified 253 such patterns. These cover the gamut from regional ("COMMUNITY OF SEVEN THOUSAND") to ornamental ("PAVING WITH CRACKS BETWEEN THE STONES") patterns—and constitute the basis for an architecture founded on organic principles that makes people feel at home.[38] Alexander's timeless way of building is the architectural equivalent of Plato's justice: the purpose of the pattern language is to order the parts so they form a harmonious and beautiful whole.

It should be possible to construct a similar pattern language for politics—a set of principles that would not dictate particular structures or institutions, which must be closely adapted to local conditions, but that would generate a political process supportive of the good life. Indeed, one of the patterns cited above is political. "COMMUNITY OF SEVEN THOUSAND" responds to a question about the optimal size of human communities that has preoccupied political philosophers since ancient times.

But let us imagine something more directly related to the political process. Suppose we were to make "PERSONAL

KNOWLEDGE REQUIRED FOR VOTING" a fundamental political principle. Although many polities could be constructed on this pattern, all of them would have to be structured so that citizens (or those acting as their agents) had direct personal knowledge of issues and candidates.

Or suppose that we were serious about preserving family farming. As we have seen, market forces tend to drive small, independent mixed farms out of business, and the subsidies intended to prevent this only abet the growth of agribusiness. Systems analysis suggests that the only way to break the perverse dynamic of "Get big, or get out" is to limit the size of farms. A pattern along the lines of "FARMS OF FORTY ACRES" might help keep farmers on the land and thereby preserve agrarian communities. Adding a pattern mandating "TENANCY IN TRUST" might assert the community's interest in responsible stewardship and help to preserve the usufruct of the land for future generations.

Only a Jeffersonian ward-republic fits these particular political-design patterns. If we used this constitutional strategy, the structure of our political institutions might not look different—after all, there is a certain logic to the executive, legislative, and judicial branches of government—but the spirit would change, perhaps radically. Yet there would also be important continuities so the wisdom of the Western political tradition would still be relevant. In reinventing the architecture of politics, we will therefore want to be guided by Plato, Rousseau, and Jefferson and also by other philosophers critical of an excessively mechanical approach to life and politics—especially Montesquieu, who was once highly influential and may yet regain his former stature.

Montesquieu's style of reasoning provides an antidote to the excessive rationalism of the Cartesian tradition. Montesquieu too seeks general laws but not at the expense of context or particulars. He says that there are "different orders of law" governing differing natural and human realms, so that "good sense consists in large measure of knowing the nuances of things."[39] Truth is therefore the product of judgment, not calculation, and rationality means taking into account all relevant factors. To put it the other way around, irrationality is not lack of logical rigor but a disregard for complexity, context, and duration—that is, for the dense web of interrelationships characteristic of social and natural systems that have evolved over time. In short, Montesquieu reasons *ecologically*. Out of this understanding of life and politics came his famous doctrine of the separation of powers, which emerged as an antidote to despotism—to the raw, unchecked, coercive power that rides roughshod over context and particulars.

Montesquieu's nuanced understanding of social reality is also evident in his political prescriptions. Like Rousseau, he understood that fallible men and women must be forced to be free: "Happy it is for men that they are in a situation in which, though their passions prompt them to be wicked, it is, nevertheless, to their interest to be humane and virtuous."[40] Like Rousseau and Jefferson and for essentially the same reasons, he favors small republics:

It is in the nature of a republic that it should have a small territory; without that, it could scarcely exist. In a large republic, there are large fortunes, and consequently little moderation of spirit; there are trusts too great to be placed in the hands of any single citizen; interests become particularized; a man begins to feel that he can become happy, great, and glorious without his country; and then, that he can become great upon the ruins of his country.

In a large republic, the common good is sacrificed to a thousand considerations; it is subordinated to various exceptions; it depends on accidents. In a small republic, the public good is more strongly felt, better known, and closer to each citizen; abuses are less extensive, and consequently less protected.[41]

Because the Western tradition has no monopoly on political wisdom Eastern thought may prove to be a vital resource as we struggle to reconstruct civilization along ecological lines. The West, says Joseph Needham, locates reality in *substance*, whereas the East finds it in *relationship*.[42] Eastern thought therefore is more in tune with ecological reality (and with the emerging world of networks) than all but a few Western philosophies. In addition, says Needham, there are two ways to progress beyond magical thinking—the Greek way of seeking *causes*, which results in a mechanical view, and the Chinese way of seeking *patterns*, which leads again to an ecological view.[43] So East and West conceive of nature in radically different ways. In the West, because nature is the creation of a celestial lawgiver, wisdom and virtue consist in discovering and obeying the structure of divine law. In the East, however, because nature emerged spontaneously—that is, it self-organized—wisdom and virtue consist in discerning and following the cosmic pattern. The basic scripture of the Taoist religion, Lao Tzu's *Tao Te Ching*, reads more like a systems primer, however mystical, than a metaphysical doctrine. It laconically sets forth the spiritual ecology of the universe and counsels humility, gentleness, simplicity, and flexibility as the way to harmonize one's thoughts and deeds with the Tao, the fundamental way of nature. Now that the Greek method has, in effect, validated the Chinese way, the wisdom of the East may help us grope our way toward a moral and political order appropriate to the age of ecology.

Whatever political institutions emerge in the future, they will be as good as the mores that undergird them—and mores do not arise, nor are they effective, without the sanction of a belief system that manifests the community's ideal. Rousseau demanded "a purely civil profession of faith . . . without which it is impossible to be a good citizen."[44] (He also proposed a natural religion, apart from this civil faith, about which more is said in the next chapter.) The Western political tradition before Hobbes is virtually unanimous on this point. As Walter Lippmann put it, without "the mandate of heaven" a polity is fated to descend into an antinomian chaos in which individuals make their own rules after the fashion of the Marquis de Sade.[45] Nor are "the rights of man," as well as other political "truths," at all "self-evident" except within a moral context. In short, civil religion will be an indispensable component of the new *politeia*.

Unfortunately, in the modern lexicon *religion* connotes the superstition and oppression of Voltaire's *infâme*, so the idea that polity must have a religious base will be stoutly resisted. The problem arises because of the type of religion prevalent in the West. To the extent that religion is defined theologically and monotheistically—that is, as the revelation of a personal God who will have no other gods before him—then a conflict with political authority is inevitable. The ultimate aim of all such religions must be a society like that of Orthodox Judaism or traditional Islam with no separation of sacred and secular.[46]

This conflict is not inevitable for the Stoic or Taoist because these forms of spirituality do not give any authority to the deity—who is in any event impersonal, who is both one and many at the same time, and who does not depend on any system of belief. Nor is this conflict inevitable for the Confucian or the Buddhist. They may urge political authority to uphold a standard of virtue—one to which most reasonable beings, including almost all Western political philosophers, would give their assent—but their motivation and rationale are pragmatic, not theological. The point is to maintain civil peace and foster social harmony for their own sake, not to carry out divine will.

Moreover, the Western phobia against religion in politics ignores the fact that, as noted in the prologue, the modern age is just as religious as the preceding one, but its secular creed—a fervent faith in progress—fails to provide a moral basis for society. Yet it does not follow from the failure of Enlightenment secularism and rationalism that we must reinstate clerical religion or an established faith. At this juncture, it is unnecessary (as well as impossible and reactionary) to resurrect old-style religious politics. As the instances above attest, particular forms of religion may be politically problematic, but not every religious tradition is politically dangerous.

We need to rediscover a sacred truth that neither conflicts with reason nor oppresses the individual and then to make that understanding the basis of a spiritualized politics. In other words, we need a nonsacerdotal, nonsectarian, nontheological, nontribal religious worldview that is compatible with science and that provides personal orientation, moral guidance, and a framework for public order without imposing dogmas that must be believed or priests who must be obeyed.[47] The eventual outcome might be a kind of Confucianism, Taoism, or Stoicism for the postmechanical age.

The ecological worldview provides the core around which such a philosophical and ethical religion could grow, for although it begins in scientific explanation, it ends in a vision that is awe-inspiring. To understand the beauty, unity, and intelligence of the cosmos is to see that it is holy; and to see that it is holy is to desire to live according to its laws—with humility, moderation, and a deep appreciation for one's connection to the totality of life on the planet. Yet no dogma, denomination, rite, or hierarchy need follow from this—only a way of living and thinking in harmony with the truth of existence, whether we call it the Good, the True, the Beautiful, or the Tao.[48]

Such a religious philosophy need not be and should not be otherworldly. The historical sociologist Pitirim Sorokin described the three basic types of culture found in civilizations—ideational, sensate, and idealistic.[49] Ideational culture is not of this world, sensate culture is solely of this world, and idealistic culture is balanced, integrating the two poles. The Enlightenment reacted to the ideational culture of the Middle Ages by going to an opposite extreme. It fostered the sensate culture of modernity, perhaps the most profane and materialistic way of life that has ever existed. To use the image of historian Henry Adams, the Virgin was supplanted by the Dynamo. Culture went from one state of imbalance to another—from belief to doubt, poetry to science, imagination to intellect, eros to logos, yin to yang.

But to be extreme or one-sided is to be pathological to some degree. Each side of a polarity has its characteristic virtues, but it also has the defects of those virtues. The price of ideational culture's community and coherence was conformity and compliance, for example. By rejecting the Virgin and embracing the Dynamo, the Enlightenment exchanged one pathology for its opposite—too much doubt, science, intellect, logos, or yang and not enough belief, poetry, imagination, eros, or yin. We gained a luxuriant freedom from conformity and compliance, but we also suffered a painful dearth of community and coherence.

Curing the political ills of our current civilization without lapsing into conformity and compliance will therefore require a conscious effort to foster cultural balance—that is, to create an idealistic civilization that integrates the two poles, instead of exalting one pole over the other. We need a marriage of the Virgin and the Dynamo.

If we succeed in creating this union, we can preserve one of the greatest achievements of modern politics—the understanding that individuals have human and civil rights that must be respected by all but especially by those in authority.[50] At the same time, we can eliminate the fatal flaw of modern politics—the self-destructive egotism that makes the individual regard himself as inherently separate from other individuals and makes the human race see itself as essentially apart from and superior to the ecological household to which it belongs. The new *politeia* will derive its authority from a spiritual and ethical philosophy that performs the essential function of religion—to reconnect us to the cosmos, thereby restoring the fundamental unity that we have lost.

Some sense of what such a philosophy might look like can be found in this excerpt from the *Western Inscription* of Zhang Zai, an eleventh-century mandarin and neo-Confucian philosopher who had the following placed on the western wall of his office, where he could see it as he performed his official duties:

Heaven is my father and Earth is my mother, and even such a small creature as I finds an intimate place in their midst. Therefore that which fills the universe I regard as my body and that which directs the universe I consider as my nature. All people are my brothers and sisters, and all things are my companions.[51]

Now that we understand what the new *politeia* entails, we might wonder anew if it is realistic to propose this radical change in our mode of politics. On the face of things, probably not, but that is not necessarily decisive. Recall Hobbes's derisory opinion of his own philosophical labor, which laid the groundwork for modern politics: "useless." I believe we have reached a critical point in human evolution that obliges us to question the basic trend of civilization since Neolithic times. We have carried the drive for material wealth and power to an extreme beyond which it cannot advance much further. It becomes daily more apparent that modern civilization confronts deep structural problems that have no plausible solution—if by solution is meant mere tinkering with our current social and political arrangements. Radical change impends, whether we seek it or not, and there is reason to believe that conditions favorable to the new *politeia* will be the final outcome—because ecological scarcity will ultimately compel us to live in smaller, simpler settings congenial to the political ideas of Rousseau and Jefferson.

Other factors will also impel decentralization—for instance, the changing nature of warfare. As John Robb points out, large, centralized systems invite asymmetric attack.[52] The resilience afforded by decentralization may also be our best protection against the global pandemics that public health officials regard as inevitable. If one should strike, localities will not be able count on much help from the center and will have to rely on their own resources. Similar concerns are now being expressed concerning the electrical grid, which is so highly interconnected that it is vulnerable to a cascade of failure. In short, decentralization looks like an idea whose time has come.

Major social change occurs in two ways—by osmosis or by upheaval. Osmosis involves a gradual and often stealthy shift in opinion until the society wakes up to a new reality: "The memorable events of history," said Le Bon, "are the visible effects of the invisible changes in human thought."[53] As we have seen, when a metaphor dies, so does its age. The *ancien régime* was doomed as soon as divine right was no longer a credible theory of politics, even though inertia kept the monarchy alive for nearly a century after it had been declared moribund.

But our own theory of politics is equally moribund. The machine metaphor inspiring our political arrangements no longer accords with the epistemology and ontology of modern science, a discontinuity that cannot continue indefinitely. In addition, the practical failures of the machine metaphor have begun to influence public debate. A small but growing minority of thinkers now understands that the exponential and thermodynamic limits to growth are real and imminent and that technology is not a panacea.

With regard to technology, I am no Luddite. There are numerous possibilities for obtaining energy from non-fossil-fuel sources—thorium reactors, geothermal wells, solar furnaces, and many others, including some yet to be invented. But there is no escape from the laws of thermodynamics. These solutions will be more costly, more challenging, and far less profitable than simply burning up found wealth. Realizing the potential of alternative fuel sources will require immense capital investments and vast energy expenditures from our rapidly dwindling stocks of fossil fuels. So we would have our work cut out for us even if we had an abundance of time. Unfortunately, we do not have time, for we have fallen behind the exponential curve. If we had had been more foresighted, we would have begun a planned transition to these alternative technologies a generation ago. Because we were not, we will suffer the consequences. Technology may eventually provide the wherewithal for an advanced civilization that is very different than our own, but it offers no quick or easy fixes to our current plight; nor will it allow us to evade the necessity for social and political change.

Fortunately, popular awareness of ecological damage and of humanity's complete dependence on nature grows by the day. Although relatively few have fully embraced the Gaian metaphor (and the consequent necessity for radical change), it has clearly begun to stake its claim on our collective consciousness. If we turn MacLeish's aphorism around to say, "A world begins when its metaphor is born," then we are witnessing the waning days of one era and the genesis of another.

The second way in which major change occurs is by upheaval, as when a gradual build-up of tectonic pressure along a geological fault suddenly releases, causing an earthquake. Inertia is a powerful force in human affairs. Through short-sightedness, wooden-headedness, selfishness, or sheer laziness, populations and regimes actively resist necessary reforms until a crisis precipitates a radical break. The moribund *ancien régime* was overthrown only after a bankrupt Louis XVI was forced to convene the Estates-General, thereby triggering the political earthquake that buried the monarchy and revolutionized France.

Although it is not possible to forecast particulars, the devastating effect of ecological scarcity on a civilization predicated on abundance will be like a slow-motion cataclysm sweeping away the old mechanical order and precipitating radical changes in our economy, society, and polity. It is unlikely that this transformation will happen according to some grand plan. Instead, we will adapt to the emerging scarcity by fits and starts as we muddle and bumble our way into the future.[54] The final outcome is predictable: a new ecological order will produce a social setting hospitable to the new *politeia*.

But the ultimate shape of the political future depends on us. The Greeks could not conceive of an alternative to the polis, even though hindsight tells us that it was moribund, and Southern slaveholders could not envision any other way of life, even though chattel slavery was economically obsolescent as well as morally indefensible. We, too, find it hard to imagine an alternative to our own constitution, even though it is obsolete both intellectually and practically.

It is human nature to cling to the devil one knows rather than embrace the angel one doesn't, so we will probably resist the required changes to the bitter end, thereby increasing the risk of a protracted and brutal transition combining the worst of Hobbes and Orwell. For the sake of posterity, let us pray that we choose instead to become conscious midwives of the Gaian age and of a *politeia* that is sane, humane, and ecological. If not, then we shall always have the consolation of philosophy. Socrates taught not that we should perfect the world but that we should perfect ourselves within an imperfect world.

To conclude, instituting *therapeia* and *paideia* will require a *politeia* that is fundamentally Rousseauean and Jeffersonian in spirit and practice. In these little republics, the individual is in direct, unmediated contact with ecological, economic, social, and political reality. Such a setting would offer to all the prospect of free and autonomous self-development within a just, moral, and fraternal community founded on natural law. The aim is not to create some apocryphal "new man" but to provide the institutional framework within which an old man—the "2,000,000-million-year-old man"—can flourish. The essential task of the new *politeia* is to discover a rule of life that reconciles nature and culture, savagery and civilization. What might be the guiding ideal of such a different mode of politics?

7
A More Experienced and Wiser Savage

Though we are not so degenerate but that we might possibly live in a cave or a wigwam or wear skins today, it certainly is better to accept the advantages, though so dearly bought, which the invention and industry of mankind offer. In such a neighborhood as this, boards and shingles, lime and bricks, are cheaper and more easily obtained than suitable caves, or whole logs, or bark in sufficient quantities, or even well-tempered clay or flat stones. . . . With a little more wit we might use these materials so as to become richer than the richest now are, and make our civilization a blessing. The civilized man is a more experienced and wiser savage.
—Henry David Thoreau[1]

Modern civilization lives on depleting energy and borrowed time. Its day of reckoning rapidly approaches. If civilization itself is to survive, it must be inspired by a new ideal that renounces endless material acquisition and makes a virtue out of the necessity of living within our ecological means. Envisioning this new civilization in any detail would require another book, but its essential spirit is succinctly stated in the epigraph above: "The civilized man is a more experienced and wiser savage."

At first glance, Thoreau's dictum seems paradoxical. Surely the whole point of civilization, at least as we normally conceive it, is to overcome and even repudiate savagery. What does it mean to be a more experienced and wiser savage? And how is this the key to making our civilization a blessing?

Other passages from *Walden* begin to resolve the paradox. After noting that the denizens of Concord seemed to be "doing penance" rather than truly living, Thoreau famously says that "the mass of men lead lives of quiet desperation."[2] They pursue "superfluities" instead of "a life of simplicity, independence, magnanimity, and trust."[3] Material simplicity, Thoreau says, is not only the cure for quiet desperation. It is also the prerequisite for spiritual abundance:

> Most of the luxuries, and many of the so called comforts of life, are not only not indispensable, but positive hinderances to the elevation of mankind. With respect to luxuries and comforts, the wisest have ever lived a more simple and meagre life than the poor. . . . None can be an impartial or wise observer of human life but from the vantage ground of what *we* should call voluntary poverty.[4]

In short, "a man is rich in proportion to the number of things he can afford to let alone."[5]

Despite his predilection for "Wildness," Thoreau does not mean for us to go back to wearing skins or living in caves and wigwams. Nostalgia for the lost primal state is futile. Civilization is here to stay and brings important material, psychological, political, moral, and even spiritual advantages. Unfortunately, we have so far mostly lacked the "wit" to create the right kind of civilization, so these advantages have been too "dearly bought." We have vainly and foolishly tried to be civilized in opposition to nature—not only to the natural world that is the matrix of human life but also to our own savage nature within. We must therefore create a civilization that transcends savagery without opposing it.

This diagnosis and prescription will not be readily accepted, for our habit of mind is to overvalue civilization and denigrate savagery. But we have much to learn and gain by becoming more experienced and wiser savages. We too easily forget that civilization is scarcely five thousand years old, that the human race broadly defined has existed for more than 99 percent of its history in savagery, that all the key inventions and innovations of civilization were made by savages, and that our savage ancestors were not brutes but scientists, philosophers, poets, and artists who created human culture. The cave paintings of Altamira and Lascaux testify that we are not more advanced than our primitive ancestors—"We have invented nothing," said Pablo Picasso after emerging from Lascaux[6]—only more powerful (and in certain respects, more "experienced"). That we do not perceive this is due to the ravages of time and to an arrogance that is only slowly being dispelled as archeologists, anthropologists, and prehistorians document the extraordinary achievements of our forebears.

To urge a more experienced and wiser savagery as the solution to the predicament of modern civilization is by no means reactionary. By calling for the rediscovery and reaffirmation of our original human nature, we can enhance it with all that we have learned by being civilized and thereby become "richer than the richest now are."

This does not require us to emulate the savage way of life, even if we now could. Our primal ancestors were not angelic. Because human nature is deeply flawed, all forms of human association are also flawed. The point is that modern civilization has caused humanity to depart so far from its natural state, both physically and psychologically, that the worst aspects of human nature predominate. Aristotle's weapons for wisdom and virtue are now employed almost exclusively for opposite ends.

The antidote to the megalomania of modern life is to be found in the savage worldview. Those whom we call savages, says anthropologist Claude Lévi-Strauss, can teach us "a lesson of modesty, decency and discretion in the face of a world that preceded our species and that will survive it."[7] Savages may be predators, but they are humble and respectful predators. They understand that they are only a small part of creation and are deeply and ineluctably connected to the cosmos, the natural world, their fellow beings, and even their prey by the bond of Eros. The loss of this understanding has made modern man uniquely dangerous and destructive. Ergo, we must resurrect Eros.

To restate this in Weberian terms, those who exist by hunting and gathering inhabit an enchanted world in which truth, beauty, goodness, and justice are seen to be immanent. In addition, nature not only imposes intrinsic limits on human greed and selfishness; it also contains inherent values opposing that greed and selfishness. The disenchantment of the world in the aftermath of the scientific and industrial revolutions removed those limits and destroyed those values, leaving humankind "free" to pursue wealth and power without scruple or constraint. As we reenchant the world by inventing a new way of life based on Eros, our savage ancestors will be our most important teachers. They incarnate the knowledge that is the essential cure for the disease of civilization—the "lesson of modesty, decency and discretion." They can show us how to ground human life once again on the natural values of humility, moderation, and connection.

At the same time, we have to acknowledge that much in our savage nature needs sublimating. Reducing a complex situation to its essence, we have not yet succeeded in the becoming truly civilized because we are still predators at heart.

The savage way of life is rooted in hunting. This does not depreciate the importance of gathering or the role of women in the primal economy. But long before the emergence of so-called patriarchal cultures, the ethos of the hunt was central to human life. Although the rise of civilization eliminated hunting as a mode of existence, it did not end predation as a way of life. Instead, civilized man now made other men his prey, seizing whole lands and peoples to feed off them.

Aristotle justified imperialism as an extension of hunting. Calling war a "natural mode of acquisition," he said "hunting ought to be practiced, not only against wild animals, but also against human beings." Just as nature has given the animals to man as lawful prey, some men "are intended by nature to be ruled by others," and wars of enslavement are therefore "naturally just."[8] The imperialism and slavery of ancient civilizations were thus a continuation of the precivilized mode of existence by other means that constituted a perversion, rather than a sublimation, of the hunting instinct.

Although it does not depend on chattel slavery, modern civilization is equally perverse, if not more so. Economic development is synonymous with plundering the storehouse of nature, and modern political economy depends on energy slavery. Modern man armed with powerful technologies is therefore a more aggressive, rapacious, and merciless predator than the savage. This fact is mostly disguised by the nature of industrial economies: remote consumers are not obliged to participate in the slaughter. Unlike our savage ancestors, we moderns are mostly unaware that we are predators and so fail to treat our prey with respect or to acknowledge the harm we do. We still live by hunting, but now we do so as heedless vandals instead of honest savages. Still before us, therefore, is the real work of civilization.

It would be premature to specify the institutional framework of a more experienced and wiser savagery. But we can grasp its essence metaphorically. If primal man was a dependent child of nature and civilized man an adolescent destroyer of nature, then the wiser savage will be a loving spouse of nature. Beyond underlining the critical importance of Eros for the future of civilization, this metaphor (along with the marriage of the Virgin and the Dynamo in the previous chapter) contains an implicit politics. Being a loving spouse implies caring, respect, mutuality, reciprocity, and interdependence.

These metaphors also point us once again to the political thought of Jefferson, whose aim was to marry the European and the Indian or the civilized and the wild, to create a way of life balanced between the extremes of primitivism and decadence, between too much of the animal and too much culture. Jefferson envisioned a politics of more experienced and wiser savagery.

So did Rousseau, whose thought is too often misinterpreted as urging a return to the primal innocence of the state of nature. But he explicitly denies that he wants humanity "to live in forests with bears."[9] It is true that men and women who were born free are now found everywhere in chains, for all the reasons that Rousseau spells out at length in *The First and Second Discourses*, but the solution is not to reject civilization itself. On the contrary, "The passage from the state of nature to the civil state produces a remarkable change in man, by substituting justice for instinct in his behavior and by giving his actions the morality they previously lacked."[10] So the civil state is necessary if men and women are to rise above savagery and become truly moral beings, but when civilization cuts itself off from nature, it becomes false, artificial, immoral, and corrupt. Because cities are the furthest from nature, they are the epitome of corruption: "Cities," said Rousseau, "are the abyss of the human species."[11]

Rousseau's solution was tripartite—political, educational, and spiritual. The first part is the best known and has already been discussed: live more simply and naturally in small, face-to-face communities rooted in the land. Rousseau did not propose this as an eminently practical solution but as an ideal for judging actual polities. He understood that a corrupt civilization full of immoral individuals would not soon reform itself.[12] Besides, said Rousseau, a merely political solution can never make men and women truly free because the real chains are not external, they are internal: "Freedom is found in no form of government; it is in the heart of the free man."[13]

So the second and more important part of Rousseau's answer is found in *Émile,* which is a manual on how to raise an urban savage:

There is a great difference between the natural man living in the state of nature and the natural man living in the state of society. Emile is not a savage to be relegated to the desert. He is a savage made to inhabit cities.[14]

Rousseau was a nature mystic who participated in the savage worldview, so the third part of his answer was spiritual. For Rousseau, the earth was a bible, and "The Profession of Faith of the Savoyard Vicar" in *Émile* is an appeal for the restoration of natural religion—a kind of purified animism—as the necessary way to reconnect human beings to the divine.[15] Like Thoreau and Jefferson, Rousseau promulgated a politics of more experienced and wiser savagery.

Thoreau, Jefferson, and Rousseau are not alone in urging us to embrace a simpler and more natural mode of existence. There is a kind of universal utopia that recurs over and over in human history. A few of the best known include Lao-Tzu's *Tao Te Ching* (especially chapter 80), Plato's *Republic* (especially book II), More's *Utopia*, Shakespeare's *Tempest* (act II, scene i),[16] Gandhi's *Hind Swaraj*, and Huxley's *Island*. All argue for a way of life that is materially and institutionally simple but culturally and spiritually rich. There seems to be an archetype of the good life lodged deep in the human psyche. We somehow know that such a life would be better, nobler, and happier, but we have so far mostly lacked the "wit" to achieve it.

It is time to put individuation together with more experienced and wiser savagery. These are not separate goals but two different ways of talking about the basic character of the required advance in civilization. As we have seen, civilization hitherto has involved a movement away from the immediacy of direct perception within a largely natural environment toward an increasingly indirect or mediated mode of perception within a predominantly artificial environment. Enlightenment epistemologists like Bacon and Descartes carried this movement away from mystical participation and toward intellectual abstraction to an extreme. They renounced and denounced participation in all its forms as a snare and a delusion, and they declared war on nature, asserting it to be a lifeless world machine to be exploited without feeling or compunction. The harm done by this refusal to connect with the rest of creation is incalculable. It has cut the human race off from the external wildness of nature and from the internal wildness of the psyche—that is, from the outer and inner wellsprings of life.

Participation involves an I-Thou relationship with creation—that is, a personal relationship characterized by immediacy and intimacy, as opposed to an impersonal, disconnected I-It relationship that keeps reality at arm's length. The savage participates in nature to such a degree that reality blazes with intrinsic meaning. Life is directly perceived to be vital and sacred, as in Blake's poetic vision of tigers burning brightly in the forests of the night. The myths of primal societies are therefore auxiliary, mere reminders of what their inhabitants experience directly, immediately, and continuously—that they are connected to all of creation and that, like Blake, they

Hold Infinity in the palm of [their] hand
And Eternity in an hour.

As humanity moves away from primal participation and toward civilized abstraction, myth becomes primary. Instead of perceiving infinity and eternity for themselves, men and women learn about them through stories and thereby enter the mythopoetic age, which is synonymous with the rise of civilization. As literacy displaces oral culture, stories give way to texts. Religion is now based on scripture, and the mythopoetic or "pagan" ways, with all of their emotional immediacy, are pushed to the margin. Truly civilized folk become "People of the Book," and those who have mastered the book—whether Analects, Bible, Koran, or Torah—take charge of the spiritual flock. Despite this further weakening of emotional immediacy, scriptural religion once sufficed to provide a vital connection to the cosmic order. Finally, however, this attenuated but still real connection was destroyed when the scientific rationality promoted by the Enlightenment, which has as its purpose the construction of knowledge, vanquished the mythic mode of understanding, which has as its purpose the construction of meaning. This lack of connection causes the spiritual vertigo of modern times.

We can now see that the Enlightenment's rejection of participation was based on a flawed epistemology. The essential lesson of systems ecology is that our fate is linked to everything else in the biosphere and that we do not and cannot exist apart from the rest of nature. The fundamental message of particle physics is that we live in a participatory cosmos in which "any local agitation shakes the whole universe." And the basic teaching of depth psychology is that our individual minds are outposts of a larger world of consciousness, so that both our nervous systems and our psyches are structured for participation.

In this light, the ultimate goal of a politics of consciousness is not merely to create better institutions. It is to transform the nature of civilization. Primal peoples of the past participated naively and unconsciously, too often lapsing into gross superstition or the kind of behavior that has made savagery practically synonymous with brutality, and the overly civilized individuals of the present refuse to participate at all, leading to the alienation and megalomania of modern times. But the individuated men and women of the future will participate once again as more experienced and wiser savages who consciously integrate the primal and civilized modes of being as well as the aesthetic and scientific modes of understanding to create a culture that is epistemologically, ecologically, and psychologically mature.

By recovering the connection with nature lost when we ceased to be savages, humanity will complete an evolutionary journey from an unconscious identification with nature, to a willful separation from nature, and finally, to a conscious reidentification with nature enriched by all that we have experienced and learned during millennia of civilized existence.

Like Hegel, I therefore see the dialectic of history moving us toward higher consciousness. This does not mean that progress toward this end will occur automatically or without pain and struggle or even temporary reversal. These things are an intrinsic part of the dialectical process. The tension between the thesis of unconscious primal participation and the antithesis of willful scientific abstraction must eventually be resolved in a synthesis—a conscious and wise participation in nature that makes the human mind whole.

The Hegelian dialectic is analogous to ecological succession. It is in the nature of social and intellectual systems, like ecosystems, to make themselves obsolete by altering the conditions that created and sustained them. In this way, successively higher stages of consciousness become necessary, moving civilization toward a climax of ever greater cultural maturity.

In the end, there is simply no alternative to a consciousness revolution. Real progress depends and has always depended on expanded consciousness, not material improvement. As Will and Ariel Durant observe, "The only real revolution is in the enlightenment of the mind and the improvement of character, the only real emancipation is individual, and the only real revolutionists are philosophers and saints."[17] But never before has humanity been confronted with the necessity for such a profound consciousness revolution. A race that has overrun the planet materially has nowhere to turn but the spiritual realm. Now that we can no longer live at the expense of the rest of creation, we must learn to live in harmony with it, and such a state of harmonious interdependence will require a mature culture that fosters inner satisfaction and intrinsic meaning.

Nor can the core political challenge of our time be met in any other way. Saving the autonomous individual from the debacle of selfish individualism will require individuation or something like it. In addition, nothing but a consciousness revolution can resurrect Eros and make it possible for human beings to reconcile their original nature with the demands of civilized existence—that is, to live as more experienced and wiser savages.

To envision the eventual shape of such a civilization is not impossible. As documented by zoologist Tim Flannery, our situation is analogous to that of the ancestors of the Aboriginal Australians and the Native Americans. When they first arrived on their respective continents, they found virgin resources in the form of vast herds of large, unwary prey animals. Gorging on this seeming cornucopia of biological wealth, their populations exploded, and they began to live as heedless consumers, believing that the riches were endless. But neither the animals nor the plants could long withstand spears and fire sticks. As overexploitation took its toll, the mega fauna were driven to extinction, and the flora radically diminished. So boom was followed by bust, precipitating a crash in the human population. The survivors were compelled to evolve ways of life that were ecologically harmonious, primarily by shifting their cultural focus away from the material to the spiritual realm.

We too are obliged to make a similar transition—to discover a new mode of civilization or to rediscover an old mode of civilization that will permit humanity to live in long-term balance with the remaining resources by redirecting our energies toward nonmaterial ends. Unfortunately, our situation is more challenging in that industrial civilization relies on finite geological capital instead of potentially renewable biological income, so any transition is likely to be more wracking, as well as more protracted.

The main unknown is how much of a technological base we can expect to retain when the energy subsidy afforded by cheap and abundant fossil fuel dwindles. Assuming that a reasonable level of technological prowess does survive the transition, it is possible to descry the essential character of ecological civilization—Bali with electronics.

Like Bali, the new civilization will of necessity be agrarian or, more specifically, horticultural because labor-intensive, biodynamic agriculture maximizes yield in ecological terms while preserving the fertility of the soil over the long-term, something the Balinese have done for centuries. Future generations should also be able to draw on a variety of renewable energy sources and sophisticated technologies to support the most useful and desirable aspects of economic development, such as modern dentistry and communications (although not everyone believes that modern communications are an unmixed blessing).[18] Most important, Bali provides an inspiring model of a possible future culture.

The Balinese are a prime example of what anthropologist Clifford Geertz calls *involution*—the skillful use of limited resources to make a culture deeper and more complex.[19] They have mimicked nature by moving toward a cultural climax in which energy is used frugally to create greater richness and refinement. Balinese society is characterized by a dense network of social relationships that both support and control individual behavior, while at the same time fostering a renowned culture.

The Balinese therefore exemplify anthropologist Ruth Benedict's recipe for a good culture to a very high degree.[20] After a lifetime of studying many different societies, Benedict concluded that cultures based on reciprocity and mutuality are "synergistic" because they tend to foster a happiness of the whole that is greater than the sum of the parts. Such cultures are objectively better than atomistic ones, which typically sanction selfishness, competition, acquisition, and aggression. An ethos of dog-eat-dog and winner-take-all produces a society that funnels power and profits to a relative few and that also engenders pervasive insecurity, even among the winners.

In addition, Balinese culture is dedicated to beauty. It is said that every Balinese is an artist. Balinese people also enjoy an abundance of leisure time with which to follow their predilections and employ their talents in pursuit of cultural ends. They alternate working in rice paddies with carving masks, making offerings, participating in rituals, playing in the gamelan, tending fighting cocks, and so on. Bali resembles Karl Marx's ideal society—one that "makes it possible for me to do one thing today and another tomorrow, to hunt in the morning, fish in the afternoon, rear cattle in the evening, criticize after dinner, just as I have a mind, without ever becoming hunter, fisherman, herdsman or critic."[21]

Finally, the Balinese village exemplifies Platonic justice. As individuals gravitate toward the roles that best suit them, each person finds his or her place in the social order and tends to bring about the harmony of the whole. Although the *banjar* or village council represents the village and oversees communal affairs, for the most part Balinese life is self-regulating. Things happen spontaneously and through unspoken consensus.

Cremation ceremonies, for example, are the apex of Balinese ritual life and require elaborate preparations, often extending over many years. There is no boss, no forced contribution, and no organization manual—yet the ceremony somehow comes together in the usual Balinese cooperative fashion. Similarly, the gamelan has no conductor, only a leader (who follows the dancer), and the players themselves select an instrument within their abilities. Those without musical talent make their contribution to village life in some other fashion.

Politically, Balinese villages are largely independent of central authority. If villagers catch a thief, they kill him. When the police show up to investigate, everyone says the same thing—"The *banjar* did it"—and the case goes into the unsolved drawer along with all the others. The authorities representing the Indonesian government have learned to keep their noses mostly out of village affairs.

The Balinese are not angels. As the treatment of thieves indicates, they can be violent,[22] and Balinese men wage proxy wars with fighting cocks.[23] But Balinese society can be a pattern for thinking about a future way of life that will need to be rooted in the soil. Other patterns are possible. The Amish are exemplary in many ways and might be a more appropriate model for American conditions, given that rice paddies and wheat fields are very different agricultural regimes. But since Bali manifests involution, synergy, beauty, leisure, and justice to a high degree, it would seem to offer an apposite model.

A political future that follows the Balinese model would require an overall authority but would have no need for cultural or religious uniformity or for top-down control. There could be a mosaic of cultures that are largely self-governing. The shape of future governance might therefore resemble the early American Articles of Confederation or the *millet* system of the Ottoman empire in which the various linguistic and religious communities were granted wide autonomy as long as they respected the suzerainty of the sultan.

Before examining the nature of that authority, we must consider the place of the city in a predominantly agrarian civilization. By definition, civilization entails cities. But cities live on the surplus of energy that they siphon away from their hinterland. When energy becomes scarce and expensive, the sprawling megacities of today will be insupportable. As the discussion in chapter 3 shows, it will never make ecological or thermodynamic sense to expend vast amounts of energy to concentrate diffuse energy and then ship it off to distant conurbations. Renewable energy resources will never provide the quantity or quality of energy necessary to sustain vast agglomerations of consumers. No amount of "greening" or densification of the modern city can alter this brute fact.

A future economy relying primarily on diffuse energy will necessarily entail a decentralized pattern of settlement resembling that of preindustrial times. Cities will be relatively small and compact, like the provincial cities or city-states of old, just as towns and villages will also be dense clusters of buildings surrounded by field and forest, like the traditional agricultural settlements of Europe and Asia.

A possible model for the urban future is to be found in Paolo Soleri's *arcologies*, a portmanteau word that merges architecture and ecology. Arcologies are engineered hyperstructures designed to house relatively large populations while minimizing their ecological impact. They are a technically sophisticated recreation of the Renaissance city-state or the antique polis, albeit in a more condensed form.

Whatever form it takes, the city of the future is bound to contain the intrinsic contradictions of urban life. On the one hand, cities are the place where morals go to die and where politics is corrupt; on the other, they are where the air makes you free and where culture in all its forms is created. The city therefore presents a more serious political challenge than the village. As the history of ancient Athens or Renaissance Florence attests, the politics of the polis or city-state can be turbulent because it is too large, complex, and divided to be governed by a *banjar*.

Once we leave behind little villages and consider a larger political setting containing cities, the political ideals of Thoreau, Jefferson, and Rousseau begin to break down. If cities require different principles of governance than villages, then so do larger collectivities of villages and cities. Little polities will always be tempted to seek advantage or even bid for supremacy—that is, to emulate the career of Rome, which began as village and ended as empire.

Some form of sovereign power to keep the peace and uphold the laws is therefore indispensable. As Hobbes teaches, without the sovereign's sword nothing can prevent an eventual descent into a nasty and brutish state of nature. Such a sovereign would necessarily obey the rules of power politics laid down by Machiavelli and Hobbes, which means that it will be subject to the usual tendency of power to become tyrannical and corrupt. In short, the larger polity cannot be governed according to the same principles as Jeffersonian ward-republics or Bali-style villages.

It might seem contradictory to combine opposing political philosophies within a single system. But as Rousseau himself said, "People have always argued about the best form of government, without considering that each of them is the best in certain cases, and the worst in others."[24] In other words, in politics as in racing, it's horses for courses. Rousseau, Jefferson, and Thoreau are good for a small, simple polity; Hobbes and Machiavelli may be the necessary guides for a larger, more complex polity. (Indeed, the American political system is now a dysfunctional morass largely because our ruling ideals and institutions, holdovers from earlier and simpler times, are inadequate for governing a globe-straddling imperial power.) As previously noted, simple-minded, ideological, one-size-fits-all solutions to complex problems are chimerical and dangerous. For civilization to survive, much less thrive, our thinking must become as sophisticated, flexible, and multifaceted as our world.

In practice, this means that any future sovereign will need to have its power tamed along the lines favored either by Locke and Madison or by Confucius and Stoics such as Seneca and Cicero. All other things being equal, the latter seems preferable. The Confucian and Stoic approaches to governance may be better adapted to what historian William McNeill sees as an inexorable trend toward the historical norm. As "civilized polyethnicity" gradually replaces "barbarian homogeneity" in the developed world, the likely political outcome is a multicultural empire.[25]

In addition, because the Confucian or Stoic notion of politics proceeds from spiritual rather than mechanical premises, it is more in keeping with the ecological ethos outlined above. Confucian political thought is especially relevant. As noted previously, the ruling metaphors of Eastern thought depict nature as an interdependent web rather than a hierarchical regime, so Eastern political philosophy starts from the premise that relationship is primordial. The challenge of politics is therefore to ensure that relationships across the board are harmonious and that the authorities are "human-hearted"—in other words, to create a well-ordered society governed by a spirit of mutuality. Confucian thought might therefore hold important lessons for a political culture that needs to redress a gross imbalance between personal rights and civic responsibilities.

Paradoxically, to follow Confucius would be to return to the root of the Western political tradition in Plato. Confucius and Plato held similar views on music, poetry, and education, and their concept of justice is virtually identical, however differently expressed: justice consists in fostering social relationships that contribute to the harmony of the whole. Like Plato, Confucius also believed that the true career of life is the perfection of one's character.

In addition, if prognosticators like Jean-Marie Guéhenno are correct in forecasting a future dominated by networks, then relationships that bypass current political boundaries and legal codes will become the new reality.[26] We will need to find ways to regulate these networks ethically and legally, a predicament roughly similar to that confronted by Confucius in ancient China.

Although the agrarian societies of the future will certainly have a shared worldview based on a definite spiritual orientation, this need not take any particular form (as the distance between the myths and rituals of the Balinese and the scripture of the Amish attests). Whatever form it does take, the worldview will necessarily embody the fundamental values of ecological polity—the values of humility, moderation, and connection mandated by nature. These in turn will entail two major political principles supportive of a more experienced and wiser savagery—frugality and fraternity.

Frugality is not the same as stinginess or asceticism. To be frugal means to be sparing in the use of resources—that is, thrifty without being pinchpenny. Frugality is the art of making as little as possible go as far as possible. The etymology of *frugal* is revealing: it comes from *frugalior*, which is derived from the Latin word for *fruit* and denotes "useful" or "worthy." It therefore fits perfectly with the essential principle of ecological economics—*usufruct*, which is the use and enjoyment of property or resources without damaging or depleting their worth so that they remain permanently useful.

In practice, frugality entails a total transformation of economic theory and practice. Instead of economizing labor and capital, as at present, we will be sparing of nature. So we will husband matter and energy with extraordinary care, using labor-intensive production allied to appropriate technology to convert limited resources into goods that possess a high use value.[27]

Frugality is not only the means of implementing moderation and simplicity. It also has an important political function. As Burke warns, only by placing voluntary chains on will and appetite are men qualified for civil liberty. Otherwise, "their passions forge their fetters." In other words, an appetitive way of life invites Leviathan. In addition, a more frugal social order is likely to be more just. According to Thales of Miletus, "If there is neither excessive wealth nor immoderate poverty in a nation, then justice may be said to prevail."[28] It is also likely to be more peaceful: "I am convinced," says Thoreau, "that if all men lived as simply as I then did, thievery and robbery would be unknown."[29] Finally, a frugal society might actually be more conducive to contentment than one in which men and women continually pursue happiness but never quite attain it. As Alexis de Tocqueville noted, the mind of the typical American is filled with "anxiety," "trepidation," and "dread" thanks to "his bootless chase of that complete felicity which forever escapes him."[30]

But the root of felicity is a beautiful way of life, not endless accumulation. Eventually, when we have recreated civilization in the name of Eros, we shall find that it takes much less to be happy than we imagined: "How simple and frugal a thing is happiness: a glass of wine, a roast chestnut, a wretched little brazier, the sound of the sea. . . . All that is required . . . is a simple, frugal heart," said Nikos Kazantzakis.[31]

The necessary complement to frugality is fraternity, the most neglected element of the modern political trinity—*liberté, égalité, fraternité*. Fraternity can be most simply defined as "brotherhood" but without sexist connotations—that is, as a feeling of belonging that unites a particular group of human beings by the bond of Eros.[32] In other words, fraternity is a recognition of social kinship that transcends biology but that nevertheless retains some of the force of "Blood is thicker than water."

The atomistic, liberal societies of today, along with their expansive notions of personal freedom, are an artifact of an abnormal and transitory period of abundance enabled by humanity's exploitation of found wealth—the virgin resources of the New World and the storehouses of untouched fossil fuels.[33] With the return of ecological scarcity, individuals will not have the same latitude to go their own way—to exist apart from or even in defiance of their community, on which they will increasingly depend for livelihood. Nothing less than a resurgence of fraternity will make the return of scarcity bearable. Without some feeling of kinship that induces us to seek or at least accept a common mode of life, the response to scarcity is likely to be Hobbesian in the worst sense—a war of all against all, ending only with the imposition of order by a heavy-handed Leviathan.

Because liberalism sees a close-knit social order as restrictive—that is, as an enemy of liberty—it overlooks or even disdains the supportive aspect of fraternity. But fraternity in one form or another is a feature of all religions and virtually all utopias, so it must respond to some deep longing in the human psyche. In addition, it is a core characteristic of all intact primal societies, so it appears to be one of the human race's most compelling archetypal needs.

Indeed, the common criticism of liberal society is that it is alienating, isolating, selfish, and lonely—that is, not synergistic in Benedict's sense. As Tocqueville famously said of American society, it "throws [every man] back forever upon himself alone, and threatens in the end to confine him entirely within the solitude of his own heart."[34] One consequence is that individuals and legal entities forcefully assert rights and shirk duties, ignoring or overriding the needs of the community as they fight for a bigger slice of the pie rather than seeking ways to improve it. The overburdened and gridlocked polity of today is the result. Fraternity would seem to be the necessary cure for the pervasive selfishness of liberal society.

In addition, a fraternal society would, by its very nature, give priority to human concerns and relationships. It would therefore demote economics from its current position as the master science of human affairs—both to respond to ecological imperatives and also to make it possible to achieve what E. F. Schumacher called "Economics As If People Mattered." This would in turn foster a politics as if people—not property, power, and profits—mattered.

In the end, only something like Ivan Illich's "conviviality"—coming together, like the Balinese, to make a festival of life—can enable us to live happily and well together once the industrial juggernaut's wheels start grinding to a halt. But what would a more frugal, fraternal, and convivial way of life imply for *égalité* and *liberté?*

With respect to equality, Thales has already provided the essential answer: material simplicity fosters justice because it imposes a rough equality of condition.[35] Despite the lip service paid to equality in modern times, rich industrial nations devoted to economic development actually manufacture inequality on an unprecedented scale, both domestically and internationally.[36] The wealthy societies of today are characterized by the radical inequality of supposed equals. By contrast,

Primal societies are characterized by the actual equality of natural unequals, both because the material basis of life cannot support significant differences of wealth and status between one individual and another and because the group will usually not allow one of its members to get too "big." An individual's reputation and influence thus depend almost entirely on personal qualities.[37]

So primal society enforces equality in terms of wealth and status, but this has the effect of promoting greater character development and individuality among its members. To revert to the Jungian term, because savages cannot pile up more wealth or seize more power, their only avenue for standing out is to become individuated—that is, to become who they truly are by developing their native personality and talents as fully as possible. Contrary to the modern view that primal peoples live rigidly traditional, taboo-ridden lives that submerge personality and crush self-expression, the anthropological literature testifies that savage life is not less human than our own. It is, in fact, in many ways more fully human.[38]

The savage version of equality would therefore seem to be superior to our brand of egalitarianism, which allows merit but discourages excellence. At the extreme, equality-as-leveling implies a Maoist cultural revolution that stamps out both difference and excellence. By contrast, primal society celebrates its successful hunters, its eloquent orators, its finest dancers, and its skillful medicine men and women—provided that they do not get "big."

Envy is a powerful force in human affairs and must be turned to ends that support, rather than destroy, individuality. What the above suggests is that the inhabitants of a more frugal and fraternal society would actually come closer to fulfilling the egalitarian ideal of a rough equality of condition while still permitting something like a natural aristocracy to emerge from the free play of personality and talent. In the end, frugality and fraternity are not opposed to genuine equality: they would actually make it possible.

With respect to *liberté*, the price of living in a more frugal and fraternal society is greater limitation on individual freedom of action. But appearances are deceiving. First, as we have seen, such a social order is intrinsically opposed not to individuality or individuation but to selfishness.

Second, although a more fraternal society would necessarily impose obligations on its members, it would also provide support—unlike our own society, which also compels people (through impersonal market forces to which they have to adapt) but otherwise leaves them "free" to sink or swim.

Third, because a frugal and fraternal society would be more equal, the liberty of some would not imply the deprivation of others. It would not be a world in which, to paraphrase Anatole France, rich and poor are equally "free" to sleep under bridges and beg in the streets.

Fourth, liberty is not license. There is no escape from Burke's dilemma. The choice is not whether to have checks on will and appetite but where and how they are to be applied. Internal self-restraint exercised by moral individuals is more conducive to civil liberty than the external compulsion of an authoritarian regime that vainly tries to fill a moral vacuum with laws and prisons. But individual morality does not arise spontaneously. The rare natural philosopher aside, it is the consequence of being steeped in the mores of a community. In this light, a frugal and fraternal society that instituted wise cultural self-limitation and pursued an uplifting vision of the good life would be a friend of liberty properly understood.

Another Hegelian movement impends. Liberty and equality are warring principles, for maximal liberty leads to minimal equality, while maximal equality entails minimal liberty (as well as maximal authority). The tension between these individualistic demands can be resolved only by the communal principle of fraternity, which synthesizes the two. Wilson Carey McWilliams begins his discussion of fraternity by stating his conclusion "that the ancients were right in seeing fraternity as a means to the ends of freedom and equality."[39]

Finally, we can no longer ignore the shadow price of our vaunted freedom: it rests on a kind of slavery—namely, energy slavery, which is the concrete manifestation of the slavery to appetite that impels us to enslave nature. Just as the wealth and leisure that allowed Plato and Aristotle to philosophize were afforded by the blighted lives of those forced to toil in Athenian mines and galleys, so too our own way of life entails the degradation or death of other beings or even of whole species and biomes—not to mention the deterioration of the natural systems that support all living things, ourselves included. Although seemingly more humane in most respects than antique slavery, a way of life based on energy slavery must in the end debase humanity as well. Illich points out the contradiction: "The energy crisis focuses concern on the scarcity of fodder for these slaves. I prefer to ask whether free men need them."[40]

Lewis Henry Morgan saw the contradiction and foresaw its resolution long ago. In *Ancient Society*, published in 1877, he said that predicating civilization on a continual expansion of "property" would eventually prove "unmanageable" and lead humanity toward "self-destruction." He therefore envisioned humanity rising above "a mere property career" to achieve "the next higher plane of society to which experience, intelligence, and knowledge are steadily tending. It will be a revival, in a higher form, of the liberty, equality, and fraternity of the ancient Gentes."[41]

Those who effect this revival will be Thoreau's more experienced and wiser savages. Civilization requires an economic surplus to underwrite specialization and leisure, the prerequisites for creating a refined culture. However, as Bali attests, obtaining this surplus need not entail the destruction of nature or the oppression of man. More experienced and wiser savages will have the wit to reconcile excellence and equity, as well as liberty and duty, and to use our technological capabilities to work with nature instead of against it to make our civilization a blessing.

Epilogue: True Liberation

When society requires to be rebuilt, there is no use in attempting to build it on the old plan.
—John Stuart Mill[1]

As far as we can discern, the sole purpose of human existence is to kindle a light in the darkness of mere being.
—C. G. Jung[2]

The Enlightenment tried to cure the five great ills of civilization with more of the same—more power, more aggression, more exploitation, more abstraction, more alienation. The result is hypercivilization—a state in which civilization's tragic flaws are amplified and intensified so that it becomes an engine of destruction, a self-devouring Moloch. The solution cannot possibly be even more of the same—hypercivilization squared, as it were. We must now invent a way of being civilized that does not repeat the errors of the past and yet embodies the wisdom of the past.

At present, we are both too civilized and not civilized enough. Too civilized: our supposed liberation from nature has made us into sick animals who foul their own nest and waste their substance in riotous living, heedless of the suffering imposed on the rest of creation and posterity. Not civilized enough: our imprudent liberation from moral restraint and social authority has made us into lonesome individuals who no longer know why they are living or what to believe or how to behave—and who therefore exist in a state of quiet desperation that leaves us exposed to every form of psychic contagion, from drug addiction to ideological zealotry. In short, the immoderate pursuit of liberation is causing civilization to founder.

The problem lies with our notion of liberation, which is perverse. We have tried to escape from natural constraints and external limits by overpowering them and from social restraints and moral checks by discarding them. In other words, we have attempted to live lawlessly. This is a fundamentally mistaken strategy for achieving either individual happiness or collective well-being.

True liberation comes from within. Rousseau epitomized this understanding when he said, "For the impulse of appetite alone is slavery, and obedience to the law one has prescribed for oneself is freedom." Ergo, real freedom is paradoxically attained only when we willingly obey some higher law.

It follows that our politics must be reconstituted to reflect this understanding. Having traveled the path of material and moral liberation to the bitter end, we are now obliged to take the path of inner mastery that liberates the human spirit. Only a politics of consciousness rooted in the moral vision of ecology can create a civilization worthy of the name— an ecological civilization in which humankind lives in harmony with nature, a conscious civilization in which men and women measure wealth in spirit not in property, and a political civilization in which "the liberty, equality, and fraternity of the ancient Gentes" can flourish once again. This would be the true liberation for which the "2,000,000-year-old man" who dwells in the depths of our being constantly longs.

Bibliographic Note

From the text and the notes, it should be obvious what I owe to Plato, Aristotle, Montesquieu, Rousseau, Burke, Jefferson, Thoreau, Le Bon, and Jung. For the most part, there is nothing more to be said. However, Jefferson and Jung require amplification because they did not write systematically about politics.

Jefferson's political career consumed much of his time and energy, and the bent of his mind was more practical than theoretical—instead of tomes on architecture and education, he produced Monticello and the University of Virginia—so he never wrote a political treatise. Instead, he scattered his political thought in a vast quantity of letters and other papers. Adrienne Koch paints a coherent picture of his philosophy, but Richard K. Matthews gets closer to Jefferson's radical spirit, and Leo Marx locates him in the larger context of a struggle between two competing visions of the American future. Jefferson is an empty political symbol today. Only a few eccentrics, like the farmer-poet Wendell Berry, take his ideas seriously, and only the Amish follow a simple, convivial agrarian way of life—as described by Berry (in *The Gift of Good*

Land), John A. Hostetler, David Kline, Donald B. Kraybill and Marc A. Olshan, and Gene Logsdon. Finally, William Morris's all-but-forgotten utopia provides an inkling of what a future Jeffersonian society might be like—a technologically sophisticated agrarian civilization animated by beauty.

Jung's *Memories, Dreams, Reflections* is a classic memoir that is also the best entry into a vast and sometimes abstruse oeuvre. For going deeper, *Psychological Reflections* is a compilation of brief excerpts that covers most aspects of Jung's thought. Then try two books written for the general public—*Modern Man in Search of a Soul* and *Man and His Symbols*—as well as *C. G. Jung Speaking*, which contains his speeches, articles, and interviews, along with remembrances by friends and associates. *The Undiscovered Self* is Jung's most political book (but see also George Czuczka and Volodymyr Odajnyk for summaries of his social and political ideas). Finally, *The Earth Has a Soul*, a collection of Jung's writings on nature, reveals him to be a more experienced and wiser savage. He was a learned, scientifically trained modern intellectual and, at the same time, an instinctive ecologist-shaman who drew his inspiration from the wildness of the natural world without and the autonomous psyche within.

Another approach to Jung is through Anthony Stevens, who both explicates and justifies him in the light of later discoveries. Jung should always be read in conjunction with Freud (especially Bruno Bettelheim's Freud). Among Jung's many followers, James Hillman is noteworthy for the way in which he locates the source of individual sickness in modern civilization's lack of either a sense of beauty or a "healing fiction." In addition, the historical works of Erich Neumann and the mythological studies of Joseph Campbell, Mircea

Eliade, Georgio di Santillana and Hertha von Dechand, and Elizabeth Sewell both support and complement Jung's approach to the psyche.

I must acknowledge those who have preceded me in trying to find a solution to the problematique of industrial civilization. Many of them reach a similar conclusion about the required answer, although they arrive at that answer by different routes. For example, Willis W. Harman sees science as an instrument for discovering what is wholesome for humankind—not just prudentially but also morally and spiritually. Along the same lines, Edward O. Wilson's *Consilience* seeks the morality implicit in our biological nature, and Roger D. Masters argues that taking biology seriously leads us back to a more naturalistic, classical conception of politics not far from that of Aristotle. Fritjof Capra is also close in spirit to my own work: he synthesizes the latest developments in physics, ecology, and systems and urges a corresponding paradigm shift, but he does not really deal with the politics of that shift. On a more practical level, Herman E. Daly and John Cobb follow Schumacher in proposing a simpler, smaller, decentralized economy built to human scale. Similarly, Wendell Berry (in *The Unsettling of America*), Gandhi (for whom see also Raghavan Iyer), Václav Havel, Ivan Illich, Leopold Kohr, and Kirkpatrick Sale all want to cut down the scale, simplify the means, and limit the speed of civilization, with the aim of making it more sane and humane. Theodore Roszak also urges a life of material simplicity and visionary abundance, and William Irwin Thompson foresees a future of metaindustrial villages—that is, Bali with electronics by another name. (Huxley's *Island* and Morris's utopia are also relevant here.) Bill McKibben offers a path to a more frugal, local, and durable future, and Warren A. Johnson describes how we can muddle our way toward it.

Turning to ecology proper, Daniel Botkin, Paul A. Colinvaux, Richard Dawkins, Stephen Jay Gould, Mahlon Hoagland and Bert Dodson, Lynn Margulis and Dorion Sagan, Thomas M. Smith and Robert Leo Smith, and Edward O. Wilson (in *The Diversity of Life*) cover the scientific bases. James Lovelock supplies the scientific ground for Gaia, and Lewis Thomas's poetic essays convey the understanding that mother earth is not a dead metaphor. Aldo Leopold's essays, published in 1949, are the classic expression of an environmental ethic. For ecological philosophy, see Bill Devall and George Sessions, Neil Evernden, John M. Meyer, George Sessions, and Donald Worster. Finally, Alfred North Whitehead's process philosophy prefigured the ecological worldview.

With regard to the human challenge of ecology, Harrison Brown was among the first to warn that industrial civilization was threatened by material limits. Donella Meadows et al.'s *Limits to Growth* and my own *Ecology and the Politics of Scarcity* established the case for ecological scarcity. William R. Catton, David Ehrenfeld, Garrett Hardin, J. R. McNeill, and Mathis Wackernagel and William Rees have solidified and extended the argument. James Lovelock and William F. Ruddiman elucidate the history, science, and risk of climate change. Fred Cottrell establishes, and Howard T. Odum and Elisabeth C. Odum describe in scientific terms, the energetic basis of life and civilization. From this basis, the latter offer a solution to the impending energy crisis in *A Prosperous Way Down*. Along the same lines, Herman E. Daly and Joshua Farley follow Nicholas Georgescu-Roegen in grounding economic theory on thermodynamics to make economic practice compatible with natural limits. John R. Ehrenfeld and James Gustave Speth criticize a flawed environmentalism and try to find a viable alternative. Stewart Brand also believes that old-style environmentalism is passé and proposes a radical hair-of-the-dog cure for our ecological ills—geoengineering. Since this is unlikely to succeed (and might well make our problems worse), a more likely scenario for the immediate future is to be found in Julia Wright's case study of how Cuba coped with an energy shock. It suggests the magnitude of the challenge for a developed economy that has much farther to fall once fossil fuels are no longer cheap and abundant. (B. H. King's description of traditional horticultural agriculture in East Asia is also apropos.)

On physics, I have included a representative selection of works by the philosophical physicists—A. S. Eddington, Werner Heisenberg, James Jeans, and Erwin Schrödinger—as well as a useful reader by Ken Wilber. From the many popular treatments, I have selected a number—Paul Davies, Freeman J. Dyson, Amit Goswami, Edward Harrison, Roger S. Jones, Michio Kaku, Eric J. Lerner, Lawrence Leshan and Henry Margenau, David Lindley, Heinz R. Pagels, Roger Penrose, and Steven Weinberg—that emphasize mostly the metaphorical, Platonic, and increasingly speculative (or even mystical) character of modern physics. On self-organization and the wisdom of systems, Gregory Bateson and Erich Jantsch are the pioneers at the macro level, and Ilya Prigogine at the micro level. Systems dynamics is clearly explained by Donella H. Meadows's *Thinking in Systems* and beautifully exemplified by Meadows et al.'s updated *Limits to Growth*. Charles J. Ryan's papers are excellent complements to Meadows, albeit somewhat more technical. On chaos and complexity,

James Gleick does an outstanding job of explaining an abstruse subject in clear language, but Mark Buchanan, Jeremy Campbell, John Gribben, Stuart A. Kauffman, Roger Lewin, and M. Mitchell Waldrop all make an important contribution to our understanding of this genuine revolution in scientific thought.

With regard to psychology, Otto Rank joins Jung in ruthlessly exposing the irrationality of an overly rational civilization, and both join Erich Fromm in urging the necessity of remaining connected to the instinctual realm, to Eros. Euripides, one of the ancient poets who inspired Freud, also speaks to this point. *The Bacchae* is as much political philosophy as psychological drama. It shows why any attempt to construct a purely rational social order is doomed to failure. For the nature of human nature, I have relied on Melvin Konner's superb summary of a vast literature, but Ernest Becker, Robin Fox, Anthony Stevens, Frans B. M. de Waal, Edward O. Wilson (in *On Human Nature*), James Q. Wilson, and Robert Wright round out the picture. (See also many of the writers on savagery cited below, as well as Stephen Budiansky, Vicki Hearne, Mary Midgley, and Irene M. Pepperberg on animal nature.) On the constitution of the human mind, William H. Calvin, Antonio R. Damasio, Howard Gardner, Daniel Goleman, R. L. Gregory, George Lakoff and Mark Johnson, Claude Lévi-Strauss (in *The Savage Mind*), Robert Ornstein, A. T. W. Simeons, and Paul Watzlawick are all useful, with Goleman's popularizations being an excellent entry point. Julian Jaynes is illuminating and important, but he needs to be read with caution and with the understanding that the ancient world was still essentially shamanic (see the commentators on Plato mentioned below).

On *paideia*, Werner Jaeger is definitive but daunting; Robert M. Hutchins may therefore be a better point of entry. Johan Huizinga's *Homo Ludens* is a useful reminder that *paideia* should not be a grim quest for answers but rather a higher form of play. If you want to know what *paideia* is not, John Taylor Gatto's short, sardonic essay on the horrors of modern-day, assembly-line "education" is both enlightening and heartbreaking. Morris Berman and Stephen E. Toulmin chronicle the development of the modern mindset: Berman sees revived participation as the necessary response to the disenchantment of the world; Toulmin's revisionist history suggests that we would have done better to ground modern science on Montaigne rather than Descartes, for this would have lead us earlier to an ecological worldview. Frederick Turner calls for a revival of the classic spirit, and O. B. Hardison Jr. argues for aesthetic education. Howard Gardner and Charles Murray propose educational reforms that take into account the different ways of being intelligent. Thomas S. Kuhn shows the critical importance of paradigms for organizing thought, and Donella H. Meadows (in *Thinking in Systems*) makes paradigms the key leverage point for changing system behavior. James Hillman's *Healing Fiction* underlines the psychological indispensability of a story that orients us in the world; Ernest Becker, Alasdair MacIntyre, and Walter Lippman (in *Public Opinion*) agree; and Daniel Quinn uses story to show how different stories lead to radically different outcomes.

The writers I have cited as the necessary background for Plato—Francis M. Cornford, Robert Earle Cushman, E. R. Dodds, Numa Denis Fustel de Coulanges, Eric A. Havelock, and Walter J. Ong—are worth reading for another reason. They reveal a very different mode of understanding and thus provide an avenue for thinking about a more experienced and wiser savagery. Likewise, Ann-ping Chin, Trevor Ling, Joseph Needham, and Alan W. Watts—and Lao Tzu—show us the more ecological mode of thought prevalent in the East.

Turning now to the savage and what he might have to offer to civilization, Claude Lévi-Strauss's *Tristes Tropiques*—part memoir, part ethnography—is philosophical anthropology at its best. Pierre Clastres is also exceptional on the character of primitive politics. Paul Shepard devoted a long and productive life to arguing for the superiority of the Pleistocene, a case also made by Daniel Quinn in story form. But Robin Clark and Geoffrey Hindley, Mark Nathan Cohen and George J. Armelagos, Jared Diamond, Stanley Diamond, Robin Fox, Jamake Highwater, Lewis Hyde, Theodora Kroeber, Dorothy Lee, Calvin Luther Martin, David Maybury-Lewis, Alan McGlashan, Lewis Henry Morgan, Marshall Sahlins, and Gordon R. Taylor all offer valuable perspectives on who the savage is and what he might have to teach us. For balance, see the more jaundiced views of Robert B. Edgerton and Lawrence H. Keeley. Participation is urged by David Abram, Owen Barfield, and Giambattista Vico (in the form of "poetic wisdom") as well a number of others cited above. Jean Liedloff focuses on the critical issue of childcare. She argues that because our young are no longer consistently held, as is typical in primal societies, they tend to grow up as unhappy quasi-orphans driven by a sense of inner deprivation. Finally, Tim Flannery shows how our primal ancestors adapted to an ecological challenge similar to our own. (Julia Wright is relevant here as well.)

On Bali, from an immense literature I have culled a very few works—the still useful ethnography by Miguel Covarrubias, as well as more recent studies by Fredrik Barth, Clifford

Geertz, and Stephen J. Lansing. To avoid being accused of an excessively utopian view, I also include Geoffrey B. Robinson's description of Bali's darker side (and see also Geertz's famous essay "Deep Play" in *The Interpretation of Cultures*).

Concerning the problematique of civilization in general, the reflections and analyses of Patricia Crone, Will and Ariel Durant, Johan Huizinga (in *The Waning of the Middle Ages*), William H. McNeill, Pitirim A. Sorokin, and Arnold J. Toynbee illuminate the past in ways that portend the future. For example, Crone sees us reverting to preindustrial times, Toynbee foresees small village-republics within a world-state, McNeill a neo-Confucian empire, and Sorokin a religious revival. Karl Polanyi's description of modern political economy's painful birth warns of the difficulty ahead as we try to create an ecological civilization. Joseph A. Tainter's theory of diminishing returns on investment in complexity suggests why all past civilizations have collapsed, while Thomas Homer-Dixon and Geoffrey Vickers discuss the particular problems, costs, and contradictions that now menace industrial civilization. (Harrison Brown and the others mentioned under ecological scarcity above are also relevant here.)

To conclude with politics, as noted in the preface I have relied on the so-called classics rather than contemporary authors to make my case because I believe these older works offer deeper insights and better answers to the challenge of ecological scarcity. However, Hannah Arendt, Walter Lippman, Alastair MacIntyre, Wilson Carey McWilliams, Michael J. Sandel, and Sheldon S. Wolin are excellent complements to my own argument. Each finds American liberalism to be philosophically incoherent and morally lacking. Along the same lines, Jacques Ellul and Langdon Winner show how untrammeled technology has captured the political process. Louis Dumont is a useful corrective to our instinctive democratic aversion to hierarchy. The idea that human beings may need benign conditioning also inspires deep resistance, but as B. F. Skinner points out, we are already thoroughly conditioned—just not in ways conducive to individual happiness or social peace. Daniel J. Boorstin and Marshall McLuhan make the same point with respect to media. They exist to brainwash us (and inevitably so, says McLuhan)—hence the necessity for a liberating *paideia*.

To close on a note of political realism, I am aware that my vision of a politics of consciousness is not the only possible outcome. Indeed, it may not even be the most likely. For all the reasons spelled out in Barrington Moore Jr., Reinhold Niebuhr, George Orwell, John Robb, and Barbara W. Tuchman (not to mention Thucydides, Tacitus, Machiavelli, and Hobbes), history may take a very different and perhaps much less benign course. It will take exceptional vision, enormous courage, and extraordinary "wit" to make a transition from the *Titanic* to a smaller, simpler, humbler vessel.

Notes

Preface

1. John Maynard Keynes, *The General Theory of Employment, Interest and Money* (London: Macmillan, 1936), 383.

Prologue

1. John Anthony Froude, cited by Lord Acton, "The Study of History," June 11, 1895 (usually misattributed to Acton himself).
2. Tim Flannery, *The Future Eaters* (New York: Grove, 2002); Tim Flannery, *The Eternal Frontier* (New York: Grove, 2002).
3. Lewis Henry Morgan, *Ancient Society* (New Brunswick, NJ: Transaction, 2000), 552.
4. Mark Nathan Cohen, *The Food Crisis in Prehistory* (New Haven: Yale University Press, 1977); Mark Nathan Cohen, *Health and the Rise of Civilization* (New Haven: Yale University Press, 1991). See also Mark Nathan Cohen and George J. Armelagos, eds., *Paleopathology at the Origins of Agriculture* (Orlando: Academic Press, 1984).
5. Jared Diamond calls the adoption of agriculture "a catastrophe from which we have never recovered" in "The Worst Mistake in the History of the Human Race," *Discover* (May 1987): 64–66.

6. Aristotle, *Politics*, trans. Ernest Barker (New York: Oxford University Press, 1995), III, ix.

7. Ibid., II, i.

Chapter 1

1. Tacitus, "Corruptissima re publica plurimae leges," *Annals*, III, 27. For an alternate translation and the full context, see Cornelius Tacitus, *The Annals*, ed. Anthony A. Barrett and trans. J. C. Yardley (New York: Oxford University Press, 2008), 109.

2. Aristotle, *The Politics*, trans. Benjamin Jowett (Mineola, NY: Dover, 2000), V, ix.

3. Hippolyte Taine, *Notes on England*, cited in Gertrude Himmelfarb, *The De-moralization of Society* (New York: Vintage, 1996), 39.

4. See Harvey A. Silvergate, *Three Felonies a Day* (New York: Encounter Books, 2009), who contends that the plethora of laws has made us all criminals.

5. Edmund Burke, "Speech to the Electors of Bristol" (1780).

6. Because Americans "are locked up for crimes ... that would rarely produce prison sentences in other countries [and] are kept incarcerated far longer than prisoners in other nations," the United States, with less than 5 percent of the world's population, "has almost a quarter of the world's prisoners." Adam Liptak, "Inmate Count in U.S. Dwarfs Other Nations'," *New York Times*, April 23, 2008. As a result, about 1 percent of the adult U.S. population is incarcerated—many for relatively minor, nonviolent offenses—and about 16 percent of state government employees work in the prison system. Adam Liptak, "One in One Hundred U.S. Adults behind Bars, New Study Says," *New York Times*, February 28, 2008.

7. For example, the Financial Crime Enforcement Network of the U.S. Treasury Department requires banks to file Suspicious Activity Reports. Because the penalty for noncompliance is severe and the criteria for suspicious activity are vague, banks report anything out of the ordinary. Similarly, teachers and therapists can be severely penalized for not reporting child abuse, so better to be suspicious than sorry.

8. Will Durant, *The Story of Civilization*, vol. 5, *The Renaissance* (New York: Simon & Schuster, 1953), 567.

9. Marquis de Sade, *Juliette*, trans. Austryn Wainhouse (New York: Grove, 1994), 605, 607.

10. Edmund Burke, "Letter to a Member of the National Assembly" (1791), *Reflections on the Revolution in France* (New York: Oxford University Press), 289.

11. John Adams, letter of October 11, 1798, cited in Robert Ferrell, *The Adams Family* (New York: Publius Press, 1969), 12.

12. Cited in Robert N. Bellah et al., *Habits of the Heart* (Berkeley: University of California Press, 1985, 254. See also Lorraine Smith Pringle, *The Political Philosophy of Benjamin Franklin* (Baltimore: Johns Hopkins University Press, 2007), 23, for Franklin's pithier version: "Only a virtuous people are capable of freedom."

13. Cicero, *De Republica*, III, xxii, 33, cited in A. P. d'Entrèves, *Natural Law* (London: Hutchinson, 1951), 25.

Chapter 2

1. Philip Slater, *Earthwalk* (Garden City, NY: Anchor, 1975), 21.

2. Diogenes Laërtius, *The Lives and Opinions of Eminent Philosophers*, cited in Robert L. Arrington, *Western Ethics* (New York: Wiley-Blackwell, 1998), 110.

3. Aristotle, *Nicomachean Ethics*, I, iii, many editions.

4. Albert Einstein, *The Einstein Reader* (New York: Citadel, 2006), 104.

5. William Ophuls, *Requiem for Modern Politics* (Boulder: Westview, 1997), 185–187.

6. Sigmund Freud, *Civilization and Its Discontents*, ed. and trans. James Strachey (New York: Norton, 1961), 26.

7. Clive Ponting, *A Green History of the World*, rev. and updated (New York: Penguin, 2007); J. Donald Hughes, *Pan's Travail* (Baltimore: Johns Hopkins University Press, 1996).

8. Ophuls, *Requiem for Modern Politics*, 150–176.

9. Jeremy Campbell, *Grammatical Man* (New York: Simon & Schuster, 1982), 264. See also James Lovelock, *The Ages of Gaia* (New York: Norton, 1988), 40, 52.

10. Ophuls, *Requiem for Modern Politics*, 158–159.

11. Ibid., 12–15 and 158–160.

12. Lynn Margulis and Dorion Sagan, *What Is Life?* (Berkeley: University of California Press, 2000), 194.

13. Richard A. Kerr, "No Longer Willful, Gaia Becomes Respectable," *Science* 240, no. 4851 (April 22, 1988): 393–395. See also Stephen H. Schneider and Penelope J. Boston, eds., *Scientists on Gaia* (Cambridge: MIT Press, 1991).

14. Lewis Thomas, *The Lives of a Cell* (New York: Penguin, 1978), 5, 145 (emphasis in original).

15. Daniel B. Botkin, *Discordant Harmonies* (New York: Oxford University Press, 1990), 151.

16. Humberto Maturana, "Neurophysiology of Cognition," in Paul L. Garvin, ed., *Cognition* (New York: Spartan Books, 1969), 7–8.

17. Cited in Leonardo Taran, *Parmenides* (Princeton: Princeton University Press, 1965), xx–xxi.

18. Bernd Heinrich, *Bumblebee Economics* (Cambridge: Harvard University Press, 2004).

19. See the works of Stephen Budiansky, *If a Lion Could Talk* (New York: Free Press, 1998); William H. Calvin, *The Cerebral Symphony* (New York: Bantam, 1989); Vicki Hearne, *Adam's Task* (New York: Knopf, 1986); Mary Midgley, *Beast and Man*, 2nd ed. (Cambridge: Cambridge University Press, 1989); Irene M. Pepperberg, *Alex and Me* (New York: Collins, 2008); and Frans B. M. de Waal, *Good Natured* (Cambridge: Harvard University Press, 1997).

20. Plato, *Timaeus*, 30D, and Marcus Aurelius, *Meditations*, VII, ix, both many editions.

21. For one poignant example, see Bernd Heinrich, "Clear-Cutting the Truth about Trees," *New York Times*, December 20, 2009: the surest way to kill a real forest is to plant trees.

Chapter 3

1. James Jeans, *The Mysterious Universe*, 2nd ed. (Cambridge: Cambridge University Press, 1931), 137.

2. William Ophuls, *Requiem for Modern Politics* (Boulder: Westview Press, 1997), 29–56.

3. Heinz Pagels, *The Cosmic Code* (New York: Bantam, 1983), 239, 274; Werner Heisenberg, *Physics and Philosophy* (New York: Harper Perennial, 2007), 107; Steven Weinberg, *Dreams of a Final Theory* (New York: Vintage, 1994), 171–172.

4. Cited in Abraham Pais, *Subtle Is the Lord* (New York: Oxford, 2005), 235.

5. Heisenberg, *Physics and Philosophy*, 58. See also David Lindley, *The End of Physics* (New York: Basic Books, 1993), 74–75.

6. Erwin Schrödinger, *What Is Life?* (Cambridge: Cambridge University Press), 121.

7. A. S. Eddington, *The Nature of the Physical World* (Whitefish, MT: Kessinger, 2005), 318.

8. Cited in Ken Wilber, *Quantum Questions*, 2nd ed. (Boston: Shambhala, 2002), 52. See also Weinberg, *Dreams of a Final Theory*, 195.

9. Cited in Wilber, *Quantum Questions*, 8.

10. Jeans, *The Mysterious Universe*, 111; James Jeans, *Physics and Philosophy* (New York: Macmillan, 1943), 16.

11. Eddington, *The Nature of the Physical World*, 276.

12. Freeman J. Dyson, *Infinite in All Directions* (New York: Harper Perennial, 2002), 297. See also William H. Calvin, *The Cerebral Symphony* (New York: Bantam, 1989), 337.

13. Richard P. Feynman, *Classic Feynman*, ed. Ralph Leighton (New York: Norton, 2005), 485.

14. Cited in Robert Bly, *News of the Universe* (San Francisco: Sierra Club Books, 1995), 38.

15. Ludwig Wittgenstein, *Tractatus Logico-Philosophicus*, trans. C. K. Ogden (New York: Cosimo Classics, 2007), para. 5.6 (emphasis deleted).

16. Francis Crick, "Thinking about the Brain," *Scientific American* 241 (1979): 219–232.

17. M. Mitchell Waldrop, *Complexity* (New York: Simon & Schuster, 1992), 102.

18. Ibid., 280.

19. Donella Meadows et al., *Limits to Growth: The Thirty-Year Update* (White River Junction, VT: Chelsea Green, 2004); William R. Catton, *Overshoot* (Champaign, IL: Illini Books, 1982). See also Mathis Wackernagel and William Rees, *Our Ecological Footprint* (Gabriola Island, BC: New Society, 1996), for an exceptionally useful way of conceiving and quantifying overshoot.

20. Erich Jantsch, *The Self-Organizing Universe* (Elmsford, NY: Pergamon, 1980), 307 and 14, 162, 164. See also Gregory Bateson, *Steps to an Ecology of Mind* (New York: Ballantine, 1972), 451, 461, 483.

21. James Gleick, *Chaos* (New York: Viking, 1987), 304.

22. Ibid., 314, quoting Joseph Ford.

23. Roger Lewin, "All for One, One for All," *New Scientist*, December 14, 1996, 28–33.

24. Stephen Jay Gould, *Wonderful Life* (New York: Norton, 1989).

25. Gleick, *Chaos*, 68.

26. Ibid., 202.

27. See Meadows et al., *Limits to Growth*, 17–49; and Donella H. Meadows, *Thinking in Systems*, ed. Diana Wright (White River Junction, VT: Chelsea Green, 2008), 58–72, for more detailed explanations with graphics.

28. However, see William Ophuls, *Ecology and the Politics of Scarcity* (San Francisco: Freeman, 1977), 107–111, for earlier warnings.

29. See Ophuls, *Ecology and the Politics of Scarcity*, 35–37, 42, 59, 71, 73–74, 92n, 107–115; and Ophuls, *Requiem for Modern Politics*, 7–11, along with the references cited, for more detail. See also Jeremy Rifkin, *Entropy* (New York: Bantam, 1981), for a popular treatment.

30. James Lovelock, *The Ages of Gaia* (New York: Norton, 1988), 23.

31. Peter M. Vitousek et al., "Human Appropriation of the Product of Photosynthesis," *Bioscience* 36 (1986): 368–373.

32. Alfred North Whitehead, *Modes of Thought* (New York: Free Press, 1968), 138.

33. *New York Post*, November 28, 1972, cited in Lawrence Leshan and Henry Margenau, *Einstein's Space and Van Gogh's Sky* (New York: Macmillan, 1982), 143.

Chapter 4

1. Carl G. Jung, *Memories, Dreams, Reflections*, ed. Aniela Jaffé and trans. Richard Winston and Clara Winston (New York: Vintage, 1989), 4.

2. Carl G. Jung, *Psychological Reflections*, ed. Jolande Jacobi (Princeton: Princeton University Press/Bollingen, 1973), 15.

3. Carl G. Jung, *C. G. Jung Speaking*, ed. William McGuire and R. F. C. Hull (Princeton: Princeton University Press, 1977), 88–90. See also Jung, *Memories*, 348. Jung anticipated many modern developments in neurology: see Daniel Goleman, *Emotional Intelligence* (New York: Bantam, 1995), 5, 9–11; Robert Ornstein, *The Right Mind* (New York: Harcourt Brace, 1997), 146–147; Anthony Stevens, *Archetypes* (New York: Quill, 1983), 261–265; Anthony Stevens, *The Two-Million-Year-Old Self* (College Station: Texas A & M University Press, 1993), 21–22, 41–42, 115.

4. Robert B. Edgerton, *Sick Societies* (New York: Free Press, 1992), 72.

5. Melvin Konner, *The Tangled Wing* (New York: Holt, Rinehart and Winston, 1982), 426.

6. Ibid., 427–428.

7. Stanley Milgram, *Obedience to Authority* (New York: HarperCollins, 1974).

8. Philip Zimbardo, *The Lucifer Effect* (New York: Random House, 2007).

9. Aristotle, *Politics*, trans. H. Rackham (Cambridge: Harvard University Press, 1944), I, i.

10. Carl G. Jung, *The Practice of Psychotherapy*, trans. Gerhard Adler and R. F. C. Hull (Princeton: Princeton University Press, 1985), 6.

11. Friedrich Nietzsche, *Beyond Good and Evil*, trans. Helen Zimmern (Radford, VA: Wilder, 2008), 56.

12. Claude Lévi-Strauss, *The Savage Mind* (Chicago: Chicago University Press, 1966).

13. Paul Watzlawick, *How Real Is Real?* (New York: Vintage, 1977), 50–54.

14. Alfred North Whitehead, *Science and the Modern World* (New York: Free Press, 1997), 51–58.

15. Because it conflicts with the democratic ethos, this statement may provoke resistance or even disbelief. However, studies have shown that only 30 to 35 percent of high school graduates attain Piaget's formal operational stage characterized by the logical use of symbols in abstract thought—that is, by scientific thinking in its most basic sense. See D. Kuhn et al., "The Development of Formal Operations in Logical and Moral Judgment," *Genetic Psychology Monographs* 95 (1977): 97–188; Alan Cramer, *Uncommon Sense* (New York: Oxford, 1993), 26, 34. This is not simply an artifact of our ineffectual educational system. As Charles Murray points out in *Real Education* (New York: Three Rivers Books, 2009), half of the population is, by definition, below average in every measurable dimension of intelligence broadly defined. Educational policies predicated on an idealistic view of human potential therefore are doomed to failure. Only the more intelligent can realistically aspire to an academic career that leads toward mastery of formal operations. In the end, says Murray, the democratic ethos would be better served by realism instead of romance so that individuals could receive the type of education best suited to their talents.

16. Richard Dawkins, *The Blind Watchmaker* (New York: Norton, 1986), xi.

17. R. L. Gregory, *Eye and Brain* (New York: McGraw-Hill, 1973), passim; Robin Fox, *The Search for Society* (New Brunswick, NJ: Rutgers University Press, 1989), 184–186; Julian Jaynes, *The Origin of Consciousness in the Breakdown of the Bicameral Mind* (Boston: Houghton Mifflin, 1982), 52–66.

18. Cited in Ralph Metzner, *The Unfolding Self* (Novato, CA: Origin Press, 1998), 10.

19. Cited in M. Mitchell Waldrop, *Complexity* (New York: Simon & Schuster, 1992), 227.

20. Cited in James Gleick, *Chaos* (New York: Viking, 1987), 262. See also Thomas S. Kuhn, *The Structure of Scientific Revolutions*, 2nd ed. (Chicago: Phoenix, 1970), about paradigms as the essential ground of scientific inquiry.

21. Cited in Gerald J. Holton, *The Advancement of Science, and Its Burdens* (New York: Cambridge University Press, 1986), 149.

22. Jaynes, *The Origin of Consciousness*, 58, 62.

23. Aristotle, *Poetics*, trans. S. H. Butcher (Mineola, NY: Dover, 1997), 47.

24. From the title poem of Muriel Rukeyser, *The Speed of Darkness* (New York: Random House, 1968).

25. Ludwig Wittgenstein, *Tractatus Logico-Philosophicus*, trans. C. K. Ogden (Mineola, NY: Dover, 1998), para. 6.52.

26. Ibid., para. 7.

27. Cited in Isaiah Berlin, *The Crooked Timber of Humanity*, ed. Henry Hardy (Princeton: Princeton University Press, 1998), 19.

28. Frans B. M. de Waal, *Good Natured* (Cambridge: Harvard University Press, 1997); Robert Wright, *The Moral Animal* (New York: Pantheon, 1994); Stevens, *Archetypes*, 220; Steven Pinker, "The Moral Instinct," *New York Times Magazine*, January 13, 2008.

29. Konner, *The Tangled Wing*, 427–428, referencing Waddington.

30. James Jeans, *Physics and Philosophy* (New York: Macmillan, 1943), 204.

31. Stevens, *Archetypes*, 282.

32. Ibid., 58–60; Stevens, *The Two-Million-Year Old Self*, 19–22.

33. Jung, *Memories*, 347. For a more technical definition, see Stevens, *Archetypes*, 296.

34. Cited in Stevens, *The Two-Million-Year-Old Self*, 7.

35. Lévi-Strauss, *The Savage Mind*; Stevens, *Archetypes*, 39–47, 220; Stevens, *The Two-Million-Year-Old Self*, 27–31; Fox, *The Search for*

Society, 20; O. B. Hardison Jr., *Entering the Maze* (New York: Oxford University Press, 1981), 277–288.

36. Fox, *The Search for Society*, 11–34; Stevens, *The Two-Million-Year Old Self*, 15–19, 65–66; Alexander J. Argyros, *A Blessed Rage for Order* (Ann Arbor: University of Michigan Press, 1992), 220–221, referencing the work of Turner and Murdock.

37. Fox, *The Search for Society*, 21–22.

38. A. T. W. Simeons, *Man's Presumptuous Brain* (New York: Dutton, 1961).

39. Jung, *Memories*, 325. See also Otto Rank, *Beyond Psychology* (New York: Dover, 1941), 37.

40. Jung, *C. G. Jung Speaking*, 89.

41. Jung, *Memories*, 302.

42. Carl G. Jung et al., *Man and His Symbols* (New York: Dell, 1968), 84.

43. Jung, *Psychological Reflections*, 232.

44. Jung, *The Practice of Psychotherapy*, 108.

45. Carl G. Jung, *Two Essays on Analytical Psychology*, trans. Gerhard Adler and R. F. C. Hull (Princeton: Princeton University Press, 1972), 155.

46. Stevens, *Archetypes*, 142. See also Erich Neumann, *Depth Psychology and a New Ethic*, trans. Eugene Rolfe (New York: Harper & Row, 1973).

47. Jung, *Two Essays*, 28.

48. See Volodymyr W. Odajnyk, *Jung and Politics* (New York: Harper & Row, 1976), 64–65, about the "compensatory externalization" of psychic contents.

49. Erich Neumann, *The Origins and History of Consciousness*, trans. R. F. C. Hull (Princeton: Princeton University Press, 1970), 389.

50. Jung, *Two Essays*, 239.

51. Jung, *Memories*, 342–343.

52. William James, *Essays in Radical Empiricism* (Mineola, NY: Dover, 2003), 112.

Chapter 5

1. Gregory Bateson, *Steps to an Ecology of Mind* (New York: Ballantine, 1972), 146; Gregory Bateson, *Mind and Nature* (New York: Bantam, 1980), 19.

2. Werner Jaeger, *Paideia*, 3 vols., trans. Gilbert Highet (New York: Oxford University Press, 1939, 1943, 1944), I, xxiii.

3. For more extended discussion of elite inevitability, see William Ophuls, *Requiem for Modern Politics* (Boulder: Westview, 1997), 203–204, 254–257, 260–263.

4. Gustave Le Bon, *The Crowd* (Mineola, NY: Dover, 2002), xiii.

5. C. G. Jung, *Psychological Reflections*, ed. Jolande Jacobi (Princeton: Princeton University Press / Bollingen, 1973), 166–167 (spelling in original).

6. James Buchan, *Frozen Desire* (New York: Farrar, Straus, Giroux, 1997), 191.

7. James Hillman, "And Huge Is Ugly," *Bloomsbury Review* (January–February 1992), 3, 20.

8. James Hillman, *Inter Views* (New York: Harper & Row, 1983), 136.

9. James Hillman, *Anima Mundi: The Return of the Soul to the World* (Woodstock, CT: Spring, 1982), 71–93 at 84.

10. James Hillman, *The Thought of the Heart*, Eranos Lectures 2 (Dallas: Spring, 1984), 29 (emphasis in the original). It follows, says Hillman, that depth psychology is also "a depth aesthetics," for the goal is to resurrect the *joie de vivre* of the patient. Ibid., 36.

11. Steven Weinberg, *Dreams of a Final Theory* (New York: Vintage, 1994), 165.

12. Christopher Alexander, *The Timeless Way of Building* (New York: Oxford University Press, 1979). See also Alain de Botton, *The Architecture of Happiness* (London: Hamish Hamilton, 2006).

13. Christopher Alexander et al., *A Pattern Language* (New York: Oxford University Press, 1977). Such universals have been found across the board. For example, despite the strange ways in which human beings sometimes disfigure themselves to be "beautiful," there

seems to be near universal agreement on what makes for a beautiful face. Sarah Kershaw, "The Sum of Your Facial Parts," *New York Times*, October 8, 2008.

14. Norman F. Carver, *Italian Hilltowns* (Kalmazoo, MI: Documan Press, 1994).

15. Robert Ornstein, *The Right Mind* (New York: Harcourt Brace, 1997); Jeremy Campbell, *Grammatical Man* (New York: Simon & Schuster, 1982), 238–253.

16. Melvin Konner, *The Tangled Wing* (New York: Holt, Rinehart and Winston, 1982), 414–418.

17. Lewis Thomas, *The Medusa and the Snail* (New York: Viking, 1979), 107.

18. Roger Penrose, *The Emperor's New Mind* (New York: Penguin, 1991), 412–413, 418.

19. Cited in Holger Kalweit, *Dreamtime and Inner Space* (Boston: Shambhala, 1988), 193.

20. Alfred North Whitehead, *Science and the Modern World* (New York: Free Press, 1997), 199, 204.

21. This is a major theme in the oeuvre of Hannah Arendt. See her *Lectures on Kant's Political Philosophy*, ed. Ronald Beiner (Chicago: University of Chicago Press, 1989); *The Human Condition*, rev. ed., ed. Margaret Canovan (Chicago: University of Chicago Press, 1998); *Between Past and Future* (New York: Penguin Classics, 2006).

22. Friedrich Schiller, *Letters on the Aesthetic Education of Man*, cited in O. B. Hardison Jr., *Entering the Maze* (New York: Oxford University Press, 1981), 74.

23. Henry David Thoreau, *Walden and Other Writings*, ed. Joseph Wood Krutch (New York: Bantam, 1962), 144.

24. C. G. Jung, *Memories, Dreams, Reflections*, ed. Aniela Jaffé and trans. Richard Winston and Clara Winston (New York: Vintage, 1989), 103.

25. Ophuls, *Requiem for Modern Politics*, 101–111, 179–191.

26. Francis Bacon, *Novem Organum* (1620), aphorism 28.

27. J. Stephen Lansing, *Priests and Programmers* (Princeton: Princeton University Press, 1991).

28. David Tenenbaum, "Traditional Drought and Uncommon Famine in the Sahel," *Whole Earth Review* (Summer 1986): 44–48.

29. For instances, see Ophuls, *Requiem for Modern Politics*, 115, 131–133, 135, 171, 174, 213, 222, 224–226, 267.

30. Donella H. Meadows, "Whole Earth Models and Systems," *CoEvolution Quarterly* (Summer 1982): 101 (emphases deleted).

31. Ibid.

32. Donella H. Meadows, "Systems Dynamics Meets the Press," *In Context* (Autumn 1989): 18.

33. Charles Murray, *Real Education* (New York: Three Rivers Books, 2009). See also note 15 in the previous chapter. The argument that social outcomes would be more equal if our educational institutions were better is belied by the experience of excellent systems elsewhere. For example, France and Japan are much better than the United States at producing high-school graduates who are literate and numerate, but both societies are also quite "elitist" in the pejorative sense. Education is not and never will be a substitute for a good politics.

34. C. G. Jung, "The Spiritual Problem of Modern Man," cited in Volodymyr W. Odajnyk, *Jung and Politics* (New York: Harper & Row, 1976), 113–114. Jung did not assert any correlation between these psychological levels and socioeconomic status, educational achievement, or cognitive complexity—indeed, he was careful to distinguish them—but some degree of overlap seems likely.

35. See Francis M. Cornford, *Principium Sapentiae* (Cambridge: Cambridge University Press, 1952); Francis M. Cornford, *From Religion to Philosophy* (Princeton: Princeton University Press, 1991); Robert Earl Cushman, *Therapeia* (New Brunswick, NJ: Transaction, 2002); E. R. Dodds, *The Greeks and the Irrational* (Berkeley: University of California Press, 2004); Numa Denis Fustel de Coulanges, *The Ancient City*, trans. Willard Small (Mineola, NY: Dover, 2006); Eric A. Havelock, *Preface to Plato* (Cambridge: Harvard University Press, 1963); Julian Jaynes, *The Origin of Consciousness in the Breakdown of the Bicameral Mind* (Boston: Houghton Mifflin, 1982); Walter J. Ong, *Orality and Literacy* (London: Methuen, 1982). They reveal an ancient Athens that was more like an encampment of Sioux Indians than a modern metropolis.

36. Plato, *Phaedo*, trans. Benjamin Jowett (Teddington, England: Echo Library, 2006), 35–36, 65.

37. To the extent that Plato could be said to have had a political program in the modern sense, *The Laws* makes a better candidate than *The Republic*. But even to say this much is probably an error. Antique philosophy, which is almost always a *quest* for truth, is fundamentally hostile to modern ideology, which is almost always an *assertion* of truth.

38. Nor am I a proponent of tyranny, of an oppressive regime like the chimerical "EcoRepublic" criticized in Val Plumwood's *Environmental Culture* (New York: Routledge, 2002), 62–74. As the next two chapters show, my aim is to establish a new moral, legal, and political *order*, not a permanent dictatorship of those with superior ecological knowledge and philosophical understanding.

39. Allan Bloom, "Interpretive Essay," in Plato, *The Republic of Plato*, trans. Allan Bloom (New York: Basic Books, 1968), 410.

40. The test of scientific truth is always pragmatic: does it allow us reliably to predict and achieve certain outcomes? Science is not a grand march toward some final, completely objective picture of reality. It is a successively greater ability to solve problems. See Thomas S. Kuhn, *The Structure of Scientific Revolutions*, 2nd ed. (Chicago: Phoenix, 1970), 160–173.

41. Percy Bysshe Shelley, *A Defense of Poetry* (1821).

42. Plato, *The Republic of Plato*, 303.

43. Howard Gardner, *Frames of Mind* (New York: Basic Books, 1993).

Chapter 6

1. Jean-Jacques Rousseau, *On the Social Contract*, ed. Roger D. Masters and trans. Judith R. Masters (New York: St. Martin's, 1978), I, vii and viii.

2. William Ophuls, *Ecology and the Politics of Scarcity* (San Francisco: Freeman, 1977).

3. Letter to Mary Gladstone, April 24, 1881, in John Dalberg-Acton, *Letters of Lord Acton to Mary Gladstone*, ed. Herbert Paul (New York: Macmillan, 1905), 196.

4. J. S. Mill, *On Liberty and Other Writings*, ed. Stefan Collini (Cambridge: Cambridge University Press, 1989), 114–115.

5. Blaise Pascal, *Pensées*, trans. Roger Ariew (Indianapolis: Hackett, 2005), 144.

6. As noted previously, even the direction of scientific research is determined by the outcome of a political struggle between rival metaphors, or paradigms. See Thomas S. Kuhn, *The Structure of Scientific Revolutions*, 2nd ed. (Chicago: Phoenix, 1970).

7. "Of the First Principles of Government," in David Hume, *Selected Essays*, ed. Stephen Copley and Andrew Edgar (New York: Oxford University Press, 2008), 24 (upper case in original).

8. "Discours sur l'Administration de la Justice Criminelle" (1767), cited in Michel Foucault, *Discipline and Punish* (New York: Pantheon, 1977), 102–103.

9. Cited in Joachim C. Fest, *The Face of the Third Reich* (New York: Pantheon, 1970), 70.

10. From "Hypocrite Auteur," in Archibald MacLeish, *Collected Poems 1917–1982* (Boston: Houghton Mifflin, 1985).

11. Donella H. Meadows, *Thinking in Systems*, ed. Diana Wright (White River Junction, VT: Chelsea Green, 2008), 145–165.

12. Meadows, *Thinking in Systems*, 164. See also Kuhn, *The Structure of Scientific Revolutions*.

13. Joseph A. Tainter, *The Collapse of Complex Societies* (New York: Cambridge University Press, 1990).

14. Rousseau, *On the Social Contract*, II, iii.

15. John Kenneth Galbraith, *The Affluent Society* (Boston: Houghton Mifflin, 1958).

16. Rousseau, *On the Social Contract*, IV, i.

17. Ibid., III, x; see also III, ii.

18. Ibid., II, xii.

19. Jean-Jacques Rousseau, *Émile, or On Education*, ed. and trans. Allan Bloom (New York: Basic Books, 1979). 85. See also Rousseau, *On the Social Contract*, 137 ed. note 31.

20. John Locke, *Two Treatises of Government*, 3rd ed., ed. Peter Laslett (New York: Cambridge University Press, 1988), 309 (emphasis in original).

21. Leo Marx, *The Machine in the Garden* (New York: Oxford University Press, 1964).

22. Thomas Jefferson, *Notes on the State of Virginia*, ed. Frank Shuffleton (New York: Penguin Classics, 1998), 170 (spelling in original).

23. Ibid., 171.

24. Letter to Horatio G. Spafford, March 17, 1814, in John P. Foley, ed., *The Jeffersonian Cyclopedia* (New York: Funk & Wagnalls 1900), 548.

25. Letter to Samuel Kercheval, July 12, 1816, in ibid., 235 (emphasis in original).

26. Letter to James Madison, September 6, 1789, in ibid., 375.

27. Letter to Joseph C. Cabell, February 2, 1816, in ibid., 393.

28. Letter to Charles Bellini, September 30, 1785, in ibid., 312.

29. Letter to Edward Carrington, January 16, 1787, in *The Papers of Thomas Jefferson*, vol. 11, ed. Julian P. Boyd et al. (Princeton: Princeton University Press, 1950–), 48–49.

30. Letter to John Adams, October 28, 1813, in Foley, *The Jeffersonian Cyclopedia*, 48–49.

31. Rousseau, *On the Social Contract*, III, v.

32. "Federalist No. 57," in Alexander Hamilton, James Madison, and John Jay, *The Federalist Papers* (Seattle: CreateSpace, 2010), 165.

33. See James L. Axtell, ed., *The Educational Writings of John Locke* (Cambridge: Cambridge University Press, 1968).

34. See Michael D. Young, *The Rise of the Meritocracy* (New Brunswick, NJ: Transaction, 1994).

35. Aristotle, *The Politics*, trans. Benjamin Jowett (Mineola, NY: Dover, 2000), VII, iv.

36. Letter to John Taylor, April 15, 1814, in *The Political Writings of John Adams*, ed. George W. Carey (Washington, DC: Gateway, 2001), 406.

37. Christopher Alexander et al., *A Pattern Language* (New York: Oxford University Press, 1977), 746–751.

38. Ibid., 70–74 and 1138–1140.

39. "In Defence of 'The Spirit of the Laws,'" cited in Sheldon S. Wolin, *The Presence of the Past* (Baltimore: Johns Hopkins University Press, 1990), 106–107. See also the rest of the chapter for an extended discussion of Montesquieu's process-oriented, ecological approach to politics.

40. Charles de Secondat, Baron de Montesquieu, *The Spirit of the Laws*, trans. Thomas Nugent and rev. J. V. Prichard. (London: Bell, 1914), XXI, xx.

41. A variant translation of ibid., VIII, 16, in Robert A. Dahl and Edward R. Tufte, *Size and Democracy* (Stanford: Stanford University Press, 1973), 7.

42. Joseph Needham, *Science and Civilisation in China*, vol. 2, *History of Scientific Thought*, corrected reprint (New York: Cambridge University Press, 1991), 78. See also Joseph Needham, "History and Human Values: A Chinese Perspective for World Science and Technology," *Centennial Review* 20, no. 1 (Winter 1976): 1–35.

43. Needham, *Science and Civilisation in China*, 165–166.

44. Rousseau, *On the Social Contract*, IV, viii.

45. Walter Lippmann, *The Public Philosophy* (New Brunswick, NJ: Transaction, 1989), 160–182.

46. Mark Lilla, *The Stillborn God* (New York: Knopf, 2007).

47. For one attempt to articulate such a worldview, see Aldous Huxley, *The Perennial Philosophy* (New York: Harper Perennial, 2009).

48. See the discussion of religion in Patricia Crone, *Pre-Industrial Societies* (Oxford: Oneworld, 2003), 123–143. Religion does not have to be scriptural, theological, or monotheistic.

49. Pitirim A. Sorokin, *Social and Cultural Dynamics*, rev. and abridged, (Boston: Porter Sargent, 1957).

50. The concept of rights attached to citizenship has deep historical roots. In ancient Rome, for example, citizens enjoyed civil liberties and legal guarantees even under the emperors. We should not make the mistake of conflating human and civil rights with representative democracy.

51. Cited in James Miller, *Chinese Religions in Contemporary Societies* (Santa Barbara: ABC-Clio, 2006), 293.

52. John Robb, *Brave New War* (New York: Wiley, 2007).

53. Gustave Le Bon, *The Crowd* (Mineola, NY: Dover, 2002), ix.

54. Warren A. Johnson, *Muddling toward Frugality*, rev. ed. (Westport, CT: Easton Studio Press, 2010).

Chapter 7

1. Henry David Thoreau, *Walden and Other Writings*, ed. Joseph Wood Krutch (New York: Bantam, 1962), 134–135.

2. Ibid., 108, 111.

3. Ibid., 116.

4. Ibid., 115 (spelling in original).

5. Ibid., 166.

6. Cited in John Lichfield, "Six Months to Save Lascaux," *The Independent* (UK), July 12, 2008. See also Gregory Curtis, *The Cave Painters* (New York: Knopf, 2007).

7. Cited in Morris Berman, *The Reenchantment of the World* (Ithaca, NY: Cornell University Press, 1981), 235.

8. Aristotle, *The Politics*, trans. Benjamin Jowett (Mineola, NY: Dover, 2000), I, viii.

9. Jean-Jacques Rousseau, *The First and Second Discourses*, ed. Roger D. Masters and trans. Roger D. and Judith R. Masters (New York: St. Martin's, 1964), 201.

10. Jean-Jacques Rousseau, *On the Social Contract*, ed. Roger D. Masters and trans. Judith R. Masters (New York: St. Martin's, 1978), I, viii.

11. Jean-Jacques Rousseau, *Émile, or On Education*, ed. and trans. Allan Bloom (New York: Basic Books, 1979), 59. Rousseau is not condemning the city per se but rather cities as they exist in a corrupt civilization.

12. Roger D. Masters, "Nothing Fails Like Success: Development and History in Rousseau's Political Teaching," *University of Ottawa Quarterly* 49, nos. 3–4 (1979): 357–376.

13. Rousseau, *Émile*, 473.

14. Ibid., 205 and see also 255.

15. Ibid., 266–313, esp. 295, 306–307.

16. See also Leo Marx, *The Machine in the Garden* (New York: Oxford University Press, 1964), 34–72.

17. Will Durant and Ariel Durant, *The Lessons of History* (New York: Simon & Schuster, 1968), 72.

18. For example, Thoreau disdained the railroad and the telegraph. See Thoreau, *Walden and Other Writings*, 173–179. And Jung doubted that technological advances had contributed to human happiness: "Mostly, they are deceptive sweetenings of existence, like speedier communications, which unpleasantly accelerate the tempo of life and leave us with less time than ever before." See his *Memories, Dreams, Reflections*, ed. Aniela Jaffé and trans. Richard Winston and Clara Winston (New York: Vintage, 1989), 236.

19. Clifford Geertz, *Agricultural Involution* (Berkeley: University of California Press, 1963). See also Arnold J. Toynbee, *A Study of History*, vols. 1–6, abridged D. C. Somervell (New York: Oxford, 1947), 198 ff about "progressive simplification."

20. Ruth Benedict, "Patterns of the Good Culture," *American Anthropologist* 72, no. 2 (1970): 320–333.

21. Karl Marx, *The German Ideology*, ed. C. J. Arthur (New York: International, 1970), 53.

22. See also Geoffrey B. Robinson, *The Dark Side of Paradise* (Ithaca, NY: Cornell University Press, 1995).

23. See "Deep Play," in Clifford Geertz, *The Interpretation of Cultures* (New York: Basic Books, 1977), 412–453.

24. Rousseau, *On the Social Contract*, III, iii.

25. William H. McNeill, *Polyethnicity and National Unity in World History* (Toronto: University of Toronto Press, 1986), 83.

26. Jean-Marie Guéhenno, *The End of the Nation-State*, trans. Victoria Elliott (Minneapolis: University of Minnesota Press, 1995). See also Alvin Toffler, *Future Shock*, 10th ed. (New York: Bantam, 1984), and *The Third Wave* (New York: Bantam, 1984).

27. Appropriate (or intermediate) technology is thermodynamically efficient and ecologically harmless and also has benign human

consequences, in both the short and long terms. See E. F. Schumacher, *Small Is Beautiful* (New York: Harper & Row, 1973).

28. Cited in Goldian VandenBroeck, *Less Is More* (Rochester, VT: Inner Traditions, 1996), 83.

29. Thoreau, *Walden and Other Writings*, 119.

30. Alexis de Tocqueville, *Democracy in America*, trans. Henry Reeves (Cambridge: Sever & Francis, 1863), II, 164.

31. Nikos Kazantzakis, *Zorba the Greek*, trans. Carl Wildman (New York: Simon & Schuster, 1996), 80.

32. Although *fraternity* has mostly masculine connotations in English, we use the term *fraternal twins* to refer to nonidentical girl twins, and there seems to be no other word adequate to the purpose, especially given its long history of usage within the Western political tradition. *Community* is weak and connotes the sharing of interests rather than sentiments.

33. See William Ophuls, *Ecology and the Politics of Scarcity* (San Francisco: Freeman, 1977), 142–145, and *Requiem for Modern Politics* (Boulder: Westview, 1997), 29–56, for more detail.

34. De Tocqueville, *Democracy in America*, II, 121.

35. See also Rousseau, *On the Social Contract*, II, xi, esp. 2nd paragraph and note thereto.

36. See Ophuls, *Requiem for Modern Politics*, 94–149, esp. pp 136–148, for more detail, as well as Rousseau's "Discourse on the Origin and Foundations of Inequality among Men" in *The First and Second Discourses*, 76–228.

37. Ophuls, *Requiem for Modern Politics*, 136.

38. See Donald W. Oliver, *Education and Community* (Berkeley: McCutcheon, 1976), 151, citing Goldenweiser. See also Jared Diamond, *Guns, Germs, and Steel* (New York: Norton, 2005); Stanley Diamond, *In Search of the Primitive* (New Brunswick, NJ: Transaction, 1974); Stanley Diamond, ed., *Primitive Views of the World* (New York: Columbia University Press, 1964); Robin Fox, *The Search for Society* (New Brunswick, NJ: Rutgers University Press, 1989); Theodora Kroeber, *Ishi in Two Worlds* (Berkeley: University of California Press, 1961); and Claude Lévi-Strauss, *The Savage Mind* (Chicago: University of Chicago Press, 1966); Claude

Lévi-Strauss, *Tristes Tropiques*, trans. John and Doreen Weightman (New York: Atheneum, 1974).

39. Wilson Carey McWilliams, *The Idea of Fraternity in America* (Berkeley: University of California Press, 1973), 7.

40. Ivan Illich, *Energy and Equity* (London: Calder and Boyars, 1974), 6.

41. Lewis Henry Morgan, *Ancient Society* (New Brunswick, NJ: Transaction, 2000), 552.

Epilogue

1. John Stuart Mill, "Coleridge," in John Stuart Mill and Jeremy Bentham, *Utilitarianism*, ed. Alan Ryan (New York: Penguin, 1987), 197.

2. C. G. Jung, *Memories, Dreams, Reflections*, ed. Aniela Jaffé and trans. Richard Winston and Clara Winston. (New York: Vintage, 1989), 326.

List of Sources

Abram, David. *The Spell of the Sensuous*. New York: Vintage, 1997.

Alexander, Christopher, et al. *A Pattern Language*. New York: Oxford University Press, 1977.

Alexander, Christopher, et al. *The Timeless Way of Building*. New York: Oxford University Press, 1979.

Arendt, Hannah. *Between Past and Future*. New York: Penguin Classics, 2006.

Arendt, Hannah. *The Human Condition*. Rev. ed. Edited by Margaret Canovan. Chicago: University of Chicago Press, 1998.

Arendt, Hannah. *Lectures on Kant's Political Philosophy*. Edited by Ronald Beiner. Chicago: University of Chicago Press, 1989.

Aristotle. *The Politics*. Translated by Benjamin Jowett. Mineola, NY: Dover, 2000.

Axtell, James L., ed. *The Educational Writings of John Locke*. Cambridge: Cambridge University Press, 1968.

Barfield, Owen. *Saving the Appearances*. New York: Harcourt, Brace & World, 1965.

Barth, Fredrik. *Balinese Worlds*. Chicago: University of Chicago Press, 1993.

Bateson, Gregory. *Mind and Nature*. New York: Bantam, 1980.

Bateson, Gregory. *Steps to an Ecology of Mind*. New York: Ballantine, 1972.

Becker, Ernest. *The Birth and Death of Meaning*. 2nd ed. New York: Free Press, 1971.

Berlin, Isaiah. *The Hedgehog and the Fox*. New York: Simon & Schuster, 1953.

Berman, Morris. *The Reenchantment of the World*. Ithaca, NY: Cornell University Press, 1981.

Berry, Wendell. *The Gift of Good Land*. San Francisco: North Point, 1983.

Berry, Wendell. *The Unsettling of America*. San Francisco: Sierra Club, 1977.

Bettelheim, Bruno. *Freud and Man's Soul*. New York: Vintage, 1984.

Boorstin, Daniel J. *The Image*. New York: Vintage, 1992.

Botkin, Daniel B. *Discordant Harmonies*. New York: Oxford University Press, 1990.

Brand, Stewart. *Whole Earth Discipline*. New York: Viking, 2009.

Brown, Harrison. *The Challenge of Man's Future*. New York: Viking, 1954.

Buchan, James. *Frozen Desire*. New York: Farrar, Straus, Giroux, 1997.

Buchanan, Mark. *Ubiquity*. New York: Three Rivers Press, 2001.

Budiansky, Stephen. *If a Lion Could Talk*. New York: Free Press, 1998.

Burke, Edmund. *Reflections on the Revolution in France*. New York: Oxford University Press, 1999.

Calvin, William H. *The Cerebral Symphony*. New York: Bantam, 1989.

Calvin, William H. *How Brains Think*. New York: Basic Books, 1996.

Campbell, Jeremy. *Grammatical Man*. New York: Simon & Schuster, 1982.

Campbell, Joseph. *The Inner Reaches of Outer Space*. 3rd ed. Novato, CA: New World Library, 2002.

Campbell, Joseph, and Johnson E. Fairchild. *Myths to Live By*. New York: Viking, 1972.

List of Sources

Capra, Fritjof. *The Turning Point.* New York: Bantam, 1984.

Capra, Fritjof. *The Web of Life.* New York: Anchor, 1997.

Catton, William R. *Overshoot.* Champaign, IL: Illini Books, 1982.

Chin, Ann-ping. *The Authentic Confucius.* New York: Scribner, 2007.

Clark, Robin, and Geoffrey Hindley. *The Challenge of the Primitives.* New York: McGraw-Hill, 1975.

Clastres, Pierre. *Society against the State.* New York: Urizen, 1977.

Cohen, Mark Nathan. *The Food Crisis in Prehistory.* New Haven: Yale University Press, 1977.

Cohen, Mark Nathan, and George J. Armelagos, eds. *Paleopathology at the Origins of Agriculture.* Orlando: Academic Press, 1984.

Colinvaux, Paul A. *Why Big Fierce Animals Are Rare.* Princeton: Princeton University Press, 1979.

Cornford, Francis M. *From Religion to Philosophy.* Princeton: Princeton University Press, 1991.

Cornford, Francis M. *Principium Sapentiae.* Cambridge: Cambridge University Press, 1952.

Cottrell, Fred. *Energy and Society.* New York: McGraw-Hill, 1955.

Covarrubias, Miguel. *Island of Bali.* New York: Knopf, 1937.

Crone, Patricia. *Pre-Industrial Societies.* Oxford: Oneworld, 2003.

Cushman, Robert Earl. *Therapeia.* New Brunswick, NJ: Transaction, 2002.

Czuczka, George. *Imprints of the Future.* Washington, DC: Daimon, 1987.

Daly, Herman E., and John Cobb. *For the Common Good.* Boston: Beacon, 1989.

Daly, Herman E., and Joshua Farley. *Ecological Economics.* 2nd ed. Washington, DC: Island Press, 2009.

Damasio, Antonio R. *Descartes' Error.* New York: Putnam's, 1994.

Davies, Paul. *Superforce.* New York: Touchstone, 1985.

Dawkins, Richard. *The Blind Watchmaker.* New York: Norton, 1986.

de Botton, Alain. *The Architecture of Happiness*. London: Hamish Hamilton, 2006.

D'Entreves, A. P. *Natural Law*. London: Hutchinson, 1951.

Devall, Bill, and George Sessions. *Deep Ecology*. Layton, UT: Gibbs Smith, 2001.

de Waal, Frans B. M. *Good Natured*. Cambridge: Harvard University Press, 1997.

Diamond, Jared. *Guns, Germs, and Steel*. New York: Norton, 2005.

Diamond, Stanley. *In Search of the Primitive*. New Brunswick, NJ: Transaction, 1974.

Diamond, Stanley, ed. *Primitive Views of the World*. New York: Columbia University Press, 1964.

di Santillana, Georgio, and Hertha von Dechend. *Hamlet's Mill*. 2nd ed. Jaffrey, NH: Godine, 1992.

Dodds, E. R. *The Greeks and the Irrational*. Berkeley: University of California Press, 2004.

Dostoyevsky, Fyodor. *The Brothers Karamazov*. Translated by Constance Garrett. New York: Signet, 1957.

Dumont, Louis. *Homo Hierarchicus*. Translated by Mark Sainsbury. Chicago: University of Chicago Press, 1970.

Durant, Will, and Ariel Durant. *The Lessons of History*. New York: Simon & Schuster, 1968.

Dyson, Freeman J. *Infinite in All Directions*. New York: Harper Perennial, 2004.

Eddington, A. S. *The Nature of the Physical World*. Whitefish, MT: Kessinger, 2005.

Edgerton, Robert B. *Sick Societies*. New York: Free Press, 1992.

Ehrenfeld, David. *The Arrogance of Humanism*. New York: Oxford University Press, 1981.

Ehrenfeld, John R. *Sustainability by Design*. New Haven: Yale University Press, 2009.

Eliade, Mircea. *The Myth of the Eternal Return*. Translated by Willard R. Trask. Princeton: Princeton University Press, 1971.

Ellul, Jacques. *The Technological Society*. Translated by John Wilkinson. New York: Vintage, 1964.

Ellul, Jacques. *The Political Illusion*. Translated by Konrad Kellen. New York: Vintage, 1972.

Ellul, Jacques. *The Technological Bluff*. Translated by Geoffrey W. Bromiley. Grand Rapids, MI: William B. Eerdmans, 1990.

Euripides. *Bacchae*. Translated Paul Woodruff. Indianapolis: Hackett, 1998.

Evernden, Neil. *The Natural Alien*. 2nd ed. Toronto: University of Toronto Press, 1993.

Evernden, Neil. *The Social Creation of Nature*. Baltimore: Johns Hopkins University Press, 1992.

Flannery, Tim. *The Eternal Frontier*. New York: Grove, 2002.

Flannery, Tim. *The Future Eaters*. New York: Grove, 2002.

Fox, Robin. *The Search for Society*. New Brunswick, NJ: Rutgers University Press, 1989.

Freud, Sigmund. *Civilization and Its Discontents*. Translated by James Strachey. New York: Norton, 1961.

Freud, Sigmund. *The Future of an Illusion*. Edited by James Strachey. New York: Anchor, 1964.

Fromm, Erich. *The Art of Loving*. New York: Bantam, 1963.

Fustel de Coulanges, Numa Denis. *The Ancient City*. Translated by Willard Small. Mineola, NY: Dover, 2006.

Galbraith, John Kenneth. *The Affluent Society*. Boston: Houghton Mifflin, 1958.

Gandhi, Mohandas. *'Hind Swaraj' and Other Writings*. Edited by Anthony J. Parel. Cambridge: Cambridge University Press, 2009.

Gardner, Howard. *Frames of Mind*. New York: Basic Books, 1993.

Gatto, John Taylor. "The Six-Lesson Schoolteacher." *Whole Earth Review* (Fall 1991): 96–99.

Geertz, Clifford. *Agricultural Involution*. Berkeley: University of California Press, 1963.

Geertz, Clifford. *The Interpretation of Cultures*. New York: Basic Books, 1977.

Georgescu-Roegen, Nicholas. *The Entropy Law and the Economic Process*. Cambridge: Harvard University Press, 1971.

Gleick, James. *Chaos*. New York: Viking, 1987.

Goleman, Daniel. *Vital Lies, Simple Truths*. New York: Simon & Schuster, 1985.

Goleman, Daniel. *Emotional Intelligence*. New York: Bantam, 1995.

Goswami, Amit. *The Self-Aware Universe*. New York: Tarcher/Putnam, 1995.

Gould, Stephen Jay. *Wonderful Life*. New York: Norton, 1989.

Gregory, R. L. *Eye and Brain*. New York: McGraw-Hill, 1973.

Gribben, John. *Deep Simplicity*. New York: Random House, 2004.

Guéhenno, Jean-Marie. *The End of the Nation-State*. Translated by Victoria Elliott. Minneapolis: University of Minnesota Press, 1995.

Hardin, Garrett. *Living within Limits*. New York: Oxford University Press, 1993.

Hardison, O. B., Jr. *Entering the Maze*. New York: Oxford University Press, 1981.

Harman, Willis W. *An Incomplete Guide to the Future*. New York: Norton, 1979.

Harrison, Edward. *Masks of the Universe*. New York: Collier, 1986.

Havel, Václav. *Summer Meditations*. Translated by Paul Wilson. New York: Knopf, 1992.

Havelock, Eric A. *Preface to Plato*. Cambridge: Harvard University Press, 1963.

Hearne, Vicki. *Adam's Task*. New York: Knopf, 1986.

Heisenberg, Werner. *Physics and Philosophy*. New York: Harper Perennial, 2007.

Highwater, Jamake. *The Primal Mind*. New York: Meridian, 1982.

Hillman, James. *Anima Mundi: The Return of the Soul to the World*. Woodstock, CT: Spring, 1982.

Hillman, James. *Healing Fiction*. Woodstock, CT: Spring, 1996.

Hillman, James. *Inter Views*. New York: Harper & Row, 1983.

Hillman, James. *The Thought of the Heart*. Eranos Lectures 2. Dallas: Spring, 1984.

Hoagland, Mahlon, and Bert Dodson. *The Way Life Works*. New York: Times Books, 1995.

Homer-Dixon, Thomas. *The Ingenuity Gap*. New York: Knopf, 2000.

Homer-Dixon, Thomas. *The Upside of Down*. Washington, DC: Island Press, 2006.

Hostetler, John A. *Amish Society*. 4th ed. Baltimore: Johns Hopkins University Press, 1993.

Huizinga, Johan. *Homo Ludens*. Boston: Beacon, 1955.

Huizinga, Johan. *The Waning of the Middle Ages*. New York: Anchor, 1954.

Hutchins, Robert M. *The Learning Society*. New York: Praeger, 1968.

Huxley, Aldous. *Island*. New York: Harper Perennial, 1972.

Huxley, Aldous. *The Perennial Philosophy*. New York: Harper Perennial, 2009.

Hyde, Lewis. *The Gift*. 2nd ed. New York: Vintage, 2007.

Illich, Ivan. *Energy and Equity*. London: Calder and Boyars, 1974.

Illich, Ivan. *Tools for Conviviality*. New York: Harper & Row, 1973.

Iyer, Raghavan. *The Moral and Political Thought of Mahatma Gandhi*. New York: Oxford University Press, 1973.

Jaeger, Werner. *Paideia*. 3 vols. Translated by Gilbert Highet. New York: Oxford University Press, 1939, 1943, 1944.

Jantsch, Erich. *The Self-Organizing Universe*. Elmsford, NY: Pergamon, 1980.

Jantsch, Erich, ed. *The Evolutionary Vision*. Boulder: Westview, 1981.

Jaynes, Julian. *The Origin of Consciousness in the Breakdown of the Bicameral Mind*. Boston: Houghton Mifflin, 1982.

Jeans, James. *The Mysterious Universe*. 2nd ed. Cambridge: Cambridge University Press, 1931.

Jeans, James. *Physics and Philosophy*. New York: Macmillan, 1943.

Jefferson, Thomas. *Notes on the State of Virginia*. Edited by Frank Shuffleton. New York: Penguin Classics, 1998.

Johnson, Warren A. *Muddling toward Frugality.* Rev. ed. Westport, CT: Easton Studio Press, 2010.

Jones, Roger S. *Physics as Metaphor.* New York: New American Library, 1983.

Jung, Carl G. *C. G. Jung Speaking.* Edited by William McGuire and R. F. C. Hull. Princeton: Princeton University Press, 1977.

Jung, Carl G. *The Earth Has a Soul.* Edited by Meredith Sabini. Berkeley: North Atlantic, 2002.

Jung, Carl G. *Memories, Dreams, Reflections.* Edited by Aniela Jaffé and translated by Richard Winston and Clara Winston. New York: Vintage, 1989.

Jung, Carl G. *Modern Man in Search of a Soul.* Translated by W. S. Dell and Cary F. Baynes. New York: Harcourt, Brace & World, 1933.

Jung, Carl G. *Psychological Reflections.* Edited by Jolande Jacobi. Princeton: Princeton/Bollingen, 1973.

Jung, Carl G. *The Undiscovered Self.* Translated by R. F. C. Hull. Princeton: Princeton/Bolligen, 1990.

Jung, Carl G., et al. *Man and His Symbols.* New York: Dell, 1968.

Kaku, Michio. *Hyperspace.* New York: Oxford University Press, 1994.

Kauffman, Stuart A. *The Origins of Order.* New York: Oxford University Press, 1993.

Keeley, Lawrence H. *War before Civilization.* New York: Oxford University Press, 1996.

King, B. H. *Farmers of Forty Centuries.* Mineola, NY: Dover, 2004.

Kline, David. *Scratching the Woodchuck.* Athens: University of Georgia Press, 1999.

Koch, Adrienne. *The Philosophy of Thomas Jefferson.* Chicago: Quadrangle, 1974.

Kohr, Leopold. *The Breakdown of Nations.* New York: Dutton, 1978.

Kohr, Leopold. *The Overdeveloped Nations.* New York: Schocken, 1978.

Konner, Melvin. *The Tangled Wing.* New York: Holt, Rinehart and Winston, 1982.

Kraybill, Donald B., and Marc A. Olshan. *The Amish Struggle with Modernity*. Lebanon, NH: University Press of New England, 1994.

Kroeber, Theodora. *Ishi in Two Worlds*. Berkeley: University of California Press, 1961.

Kuhn, Thomas S. *The Structure of Scientific Revolutions*. 2nd ed. Chicago: Phoenix, 1970.

Lakoff, George, and Mark Johnson. *Metaphors We Live By*. Chicago: University of Chicago Press, 1980.

Lansing, J. Stephen, *Perfect Order*. Princeton: Princeton University Press, 2006.

Lansing, J. Stephen. *Priests and Programmers*. Princeton: Princeton University Press, 1991.

Lao Tzu. *Tao Te Ching*. Translated by Arthur Waley. Ware, England: Wordsworth, 1997.

Le Bon, Gustave. *The Crowd*. Mineola, NY: Dover, 2002.

Lee, Dorothy. *Freedom and Culture*. Englewood Cliffs, NJ: Spectrum, 1959.

Leopold, Aldo. *A Sand County Almanac*. New York: Oxford University Press, 2001.

Lerner, Eric J. *The Big Bang Never Happened*. New York: Vintage, 1992.

Leshan, Lawrence, and Henry Margenau. *Einstein's Space and Van Gogh's Sky*. New York: Macmillan, 1982.

Lévi-Strauss, Claude. *The Savage Mind*. Chicago: University of Chicago Press, 1966.

Lévi-Strauss, Claude. *Tristes Tropiques*. Translated by John and Doreen Weightman. New York: Atheneum, 1974.

Lewin, Roger. *Complexity*. New York: Macmillan, 1992.

Liedloff, Jean. *The Continuum Concept*. New York: Knopf, 1977.

Lindley, David. *The End of Physics*. New York: Basic Books, 1993.

Ling, Trevor. *The Buddha*. New York: Penguin, 1976.

Lippmann, Walter. *The Good Society*. New Brunswick, NJ: Transaction, 2004.

Lippmann, Walter. *Public Opinion*. New York: Free Press, 1966.

Lippmann, Walter. *The Public Philosophy*. New Brunswick, NJ: Transaction, 1989.

Logsdon, Gene. *Living at Nature's Pace*. Rev. ed. White River Junction, VT: Chelsea Green, 2000.

Lovelock, James. *The Ages of Gaia*. New York: Norton, 1988.

Lovelock, James. *Healing Gaia*. New York: Harmony Books, 1991.

Lovelock, James. *The Vanishing Face of Gaia*. New York: Basic Books, 2009.

MacIntyre, Alasdair. *After Virtue*. 3rd ed. Notre Dame, IN: University of Notre Dame Press, 2007.

Margulis, Lynn, and Dorion Sagan. *Microcosmos*. Berkeley: University of California Press, 1997.

Margulis, Lynn, and Dorion Sagan. *What Is Life?* Berkeley: University of California Press, 2000.

Martin, Calvin Luther. *In the Spirit of the Earth*. Baltimore: Johns Hopkins University Press, 1992.

Marx, Leo. *The Machine in the Garden*. New York: Oxford University Press, 1964.

Masters, Roger D. *Beyond Relativism*. Hanover, NH: University Press of New England, 1993.

Masters, Roger D. *The Nature of Politics*. New Haven: Yale University Press, 1991.

Matthews, Richard K. *The Radical Politics of Thomas Jefferson*. Lawrence: University of Kansas Press, 1984.

Maybury-Lewis, David. *Millennium*. New York: Viking Penguin, 1992.

McGlashan, Alan. *The Savage and Beautiful Country*. Boston: Houghton Mifflin, 1967.

McKibben, Bill. *Deep Economy*. New York: Henry Holt, 2007.

McLuhan, Marshall. *Understanding Media*. New York: McGraw-Hill, 1964.

McNeill, J. R. *Something New under the Sun*. New York: Norton, 2000.

McNeill, William H. *Polyethnicity and National Unity in World History*. Toronto: University of Toronto Press, 1986.

McWilliams, Wilson Carey. *The Idea of Fraternity in America*. Berkeley: University of California Press, 1973.

Meadows, Donella H. *Thinking in Systems*. Edited by Diana Wright. White River Junction, VT: Chelsea Green, 2008.

Meadows, Donella, Jorgen Randers, and Dennis Meadows. *Limits to Growth: The Thirty-Year Update*. White River Junction, VT: Chelsea Green, 2004.

Meyer, John M. *Political Nature*. Cambridge: MIT Press, 2001.

Midgley, Mary. *Beast and Man*. 2nd ed. New York: Routledge, 1995.

Mill, J. S. *On Liberty and Other Writings*. Edited by Stefan Collini. Cambridge: Cambridge University Press, 1989.

Montaigne, Michel Eyquem de. *The Complete Essays*. Translated by Donald M. Frame. Stanford: Stanford University Press, 1965.

Montesquieu, Charles de Secondat, Baron de. *The Spirit of the Laws*. Translated by Thomas Nugent and revised by J. V. Prichard. London: Bell, 1914.

Moore, Barrington, Jr. *Reflections on the Causes of Human Misery and upon Certain Proposals to Eliminate Them*. Boston: Beacon, 1972.

Morgan, Lewis Henry. *Ancient Society*. New Brunswick, NJ: Transaction, 2000.

Morris, William. *News from Nowhere*. Edited by David Leopold. New York: Oxford University Press, 2009.

Murray, Charles. *Real Education*. New York: Three Rivers Books, 2009.

Needham, Joseph. "History and Human Values: A Chinese Perspective for World Science and Technology." *Centennial Review* 20 (1) (Winter 1976): 1–35.

Needham, Joseph. Science and Civilisation in China. Vol. 2, *History of Scientific Thought*, corrected reprint. New York: Cambridge University Press, 1991.

Neumann, Erich. *Depth Psychology and a New Ethic*. Translated by Eugene Rolfe. New York: Harper & Row, 1973.

Neumann, Erich. *The Origins and History of Consciousness*. Translated by R. F. C. Hull. Princeton: Princeton University Press, 1970.

Niebuhr, Reinhold. *Moral Man and Immoral Society*. New York: Charles Scribner's Sons, 1932.

Odajnyk, Volodymyr W. *Jung and Politics*. New York: Harper & Row, 1976.

Odum, Howard T. *Ecological and General Systems*. Rev. ed. Boulder: University Press of Colorado, 1994.

Odum, Howard T. *Environment, Power and Society for the Twenty-first Century*. New York: Columbia University Press, 2007.

Odum, Howard T., and Elisabeth C. Odum. *Energy Basis for Man and Nature*. 2nd ed. New York: McGraw-Hill, 1981.

Odum, Howard T., and Elisabeth C. Odum. *A Prosperous Way Down*. Boulder: University Press of Colorado, 2008.

Ong, Walter J. *Orality and Literacy*. London: Methuen, 1982.

Ophuls, William. *Ecology and the Politics of Scarcity*. San Francisco: Freeman, 1977.

Ophuls, William. *Requiem for Modern Politics*. Boulder: Westview, 1997.

Ophuls, William, and A. Stephen Boyan Jr. *Ecology and the Politics of Scarcity Revisited*. New York: Freeman, 1992.

Ornstein, Robert. *The Evolution of Consciousness*. New York: Simon & Schuster, 1991.

Ornstein, Robert. *The Right Mind*. New York: Harcourt Brace, 1997.

Orwell, George. *Animal Farm and 1984*. Boston: Houghton Mifflin Harcourt, 2003.

Pagels, Heinz R. *The Cosmic Code*. New York: Bantam, 1983.

Penrose, Roger. *The Emperor's New Mind*. New York: Penguin, 1991.

Pepperberg, Irene M. *Alex and Me*. New York: Collins, 2008.

Plato. *Phaedo*. Translated by Benjamin Jowett. Teddington, England: Echo Library, 2006.

Plato. *The Republic of Plato*. Translated by Allan Bloom. New York: Basic Books, 1968.

Polanyi, Karl. *The Great Transformation*. Boston: Beacon, 1957.

Prigogine, Ilya, and Isabelle Stengers. *Order out of Chaos*. New York: Bantam, 1984.

Quinn, Daniel. *Ishmael*. New York: Bantam, 1992.

Rank, Otto. *Beyond Psychology*. New York: Dover, 1941.

Rifkin, Jeremy. *Entropy*. New York: Bantam, 1981.

Robb, John. *Brave New War*. New York: Wiley, 2007.

Robinson, Geoffrey B. *The Dark Side of Paradise*. Ithaca, NY: Cornell University Press, 1995.

Roszak, Theodore. *Unfinished Animal*. New York: HarperCollins, 1977.

Roszak, Theodore. *Person/Planet*. New York: Anchor, 1978.

Roszak, Theodore. *The Voice of the Earth*. New York: Simon & Schuster, 1992.

Rousseau, Jean-Jacques. *Émile, or On Education*. Edited and translated by Allan Bloom. New York: Basic Books, 1979.

Rousseau, Jean-Jacques. *The First and Second Discourses*. Edited by Roger D. Masters and translated by Roger D. Masters and Judith R. Masters. New York: St. Martin's, 1964.

Rousseau, Jean-Jacques. *On the Social Contract*. Edited by Roger D. Masters and translated by Judith R. Masters. New York: St. Martin's, 1978.

Ruddiman, William F. *Plows, Plagues, and Petroleum*. Princeton: Princeton University Press, 2005.

Ryan, Charles J. "The Choices in the Next Energy and Social Revolution." *Technological Forecasting and Social Change* 16 (1980): 191–208.

Ryan, Charles J. "The Overdeveloped Society." *Stanford Magazine* (Fall–Winter 1979): 58–65.

Sahlins, Marshall. *Culture and Practical Reason*. Chicago: Chicago University Press, 1976.

Sahlins, Marshall. *Stone-Age Economics*. Chicago: Aldine-Atherton, 1970.

Sale, Kirkpatrick. *Dwellers in the Land*. San Francisco: Sierra Club, 1985.

Sale, Kirkpatrick. *Human Scale*. New York: Coward, McCann & Geoghegan, 1980.

Sandel, Michael J. *Democracy's Discontent*. Cambridge: Belknap Press, 1998.

Schneider, Stephen H., and Penelope J. Boston, eds. *Scientists on Gaia*. Cambridge: MIT Press, 1991.

Schrödinger, Erwin. *Mind and Matter*. Cambridge: Cambridge University Press, 1959.

Schrödinger, Erwin. *What Is Life?* Cambridge: Cambridge University Press, 1992.

Schumacher, E. F. *Small Is Beautiful*. New York: Harper & Row, 1973.

Seneca. *Letters from a Stoic*. Translated by Robin Campbell. New York: Penguin, 1969.

Sessions, George, ed. *Deep Ecology for the Twenty-first Century*. Boston: Shambhala, 1995.

Sewell, Elizabeth. *The Orphic Voice*. London: Routledge & Kegan Paul, 1955.

Shepard, Paul. *Coming Home to the Pleistocene*. Edited by Florence R. Shepard. Washington, DC: Island Press, 2004.

Shepard, Paul. *Nature and Madness*. San Francisco: Sierra Club Books, 1982.

Shepard, Paul. *The Tender Carnivore and the Sacred Game*. New York: Scribner's, 1973.

Simeons, A. T. W. *Man's Presumptuous Brain*. New York: Dutton, 1961.

Skinner, B. F. *Beyond Freedom and Dignity*. Indianapolis: Hackett, 2002.

Skinner, B. F. *Walden Two*. Indianapolis: Hackett, 2005.

Smith, Thomas M., and Robert Leo Smith. *Elements of Ecology*. 7th ed. Upper Saddle River, NJ: Benjamin Cummings, 2008.

Soleri, Paolo. *Arcology*. Cambridge: MIT Press, 1969.

Soleri, Paolo. *The Urban Ideal*. Edited by John Strohmeier. Berkeley: Berkeley Hills Books, 2001.

Sorokin, Pitirim A. *The Crisis of Our Age*. New York: Dutton, 1944.

Sorokin, Pitirim A. *Social and Cultural Dynamics*. Rev. and abridged. Boston: Porter Sargent, 1957.

Speth, James Gustave. *The Bridge at the End of the World*. New Haven: Yale University Press, 2008.

Stevens, Anthony. *Archetypes*. New York: Quill, 1983.

Stevens, Anthony. *The Two Million-Year-Old Self*. College Station: Texas A & M University Press, 1993.

Tainter, Joseph A. *The Collapse of Complex Societies*. New York: Cambridge University Press, 1990.

Taylor, Gordon R. *Rethink*. New York: Dutton, 1973.

Thomas, Lewis. *The Lives of a Cell*. New York: Penguin, 1978.

Thomas, Lewis. *The Medusa and the Snail*. New York: Viking, 1979.

Thompson, William Irwin. *Darkness and Scattered Light*. New York: Anchor, 1978.

Thompson, William Irwin. *Passages about Earth*. New York: Harper & Row, 1974.

Thoreau, Henry David. *Walden and Other Writings*. Edited by Joseph Wood Krutch. New York: Bantam, 1962.

Toffler, Alvin. *Future Shock*. 10th ed. New York: Bantam, 1984.

Toffler, Alvin. *The Third Wave*. New York: Bantam, 1984.

Toulmin, Stephen E. *Cosmopolis*. Chicago: University of Chicago Press, 1992.

Toynbee, Arnold J. *Mankind and Mother Earth*. New York: Oxford University Press, 1976.

Tuchman, Barbara W. *The March of Folly*. New York: Knopf, 1984.

Turner, Frederick. *The Culture of Hope*. New York: Free Press, 1995.

Vickers, Geoffrey. *Freedom in a Rocking Boat*. Harmondsworth, England: Penguin, 1972.

Vico, Giambattista. *New Science*. 3rd ed. Translated by David Marsh. New York: Penguin, 2000.

Wackernagel, Mathis, and William Rees. *Our Ecological Footprint*. Gabriola Island, BC: New Society, 1996.

Waldrop, M. Mitchell. *Complexity*. New York: Simon & Schuster, 1992.

Watts, Alan W. *Nature, Man, and Woman*. New York: Vintage, 1970.

Watzlawick, Paul. *How Real Is Real?* New York: Vintage, 1977.

Weinberg, Steven. *Dreams of a Final Theory*. New York: Vintage, 1994.

Whitehead, Alfred North. *Modes of Thought*. New York: Free Press, 1968.

Whitehead, Alfred North. *Science and the Modern World*. New York: Free Press, 1997.

Wilber, Ken. *Quantum Questions*. Rev. ed. Boston: Shambhala, 2002.

Wilson, Edward O. *Consilience*. New York: Vintage, 1999.

Wilson, Edward O. *The Diversity of Life*. Cambridge: Harvard University Press, 1982.

Wilson, Edward O. *On Human Nature*. Rev. ed. Cambridge: Harvard University Press, 2004.

Wilson, James Q. *The Moral Sense*. New York: Free Press, 1993.

Winner, Langdon. *Autonomous Technology*. Cambridge: MIT Press, 1977.

Wolin, Sheldon S. *The Presence of the Past*. Baltimore: Johns Hopkins University Press, 1990.

Worster, Donald. *Nature's Economy*. 2nd ed. New York: Cambridge University Press, 1994.

Wright, Julia. *Sustainable Agriculture and Food Security in an Era of Oil Scarcity*. London: Earthscan, 2009.

Wright, Robert. *The Moral Animal*. New York: Pantheon, 1994.

Young, Michael D. *The Rise of the Meritocracy*. New Brunswick, NJ: Transaction, 1994.

Index

Acton, Lord (John Dalberg-Acton), 3, 130
Adams, Henry, 159
Adams, John, 150, 188
Aesthetic unity, 97, 99, 102, 104. *See also* Beauty
Africa, 111–112
Alexander, Christopher, 104, 153
Altamira, 169
Amish, 181, 186
Ancient Greece, 117–118, 221n35
Appropriate technology, 186, 227n27. *See also* Technology
Archetypes. *See* Jung, Carl G.
Architecture, xiv. *See also* Alexander, Christopher
Arcologies, 182–183
Aristotle, 192
 and character, 149
 and human nature, 72, 82, 170
 and hunting, 171
 and knowledge, 26–27
 and metaphor, 78, 105
 and mores, 13
 and rule of life, 6, 22, 98
Arthur, Brian, 77
Articles of Confederation, 187
Ataturk, Mustafa Kemal, 132

Bacon, Francis, 28, 31, 109, 174
Balance, 29, 32–33, 36–37, 60, 67, 93, 109, 179
Bali, 179–182, 184, 186, 189, 193
 agriculture in, 110–111
 beauty in, 180–181
 justice in, 181
Bateson, Gregory, 55, 97–98, 110, 116
Beauty, 101–105, 107–108, 120, 180–181
Benedict, Ruth, 180, 188
Berlin, Isaiah, 80
Bettelheim, Bruno, 84
Blake, William, 175
Bloom, Allan, 119
Botkin, Daniel B., 37
Buchan, James, 101

Buddhism, 118, 158
Burke, Edmund, 18, 41, 72, 132, 145, 186, 191

Campbell, Jeremy, 30
Capital stocks, 65–68, 109, 138, 163, 179. *See also* Income flows
Chaos theory, 56–60
Churchill, Winston, xiv
Cicero, 20–21, 184
City, 137, 173, 182–183
Civil religion. *See* Religion
Civil rights. *See* Liberty
Civilization, ills of, 1–2, 8, 195
Climate change, 55, 58, 61, 73, 141
Climax ecosystem, 31–32, 35, 42, 65–67. *See also* Pioneer ecosystem
 as model for politics, 32, 42, 67, 177, 179–180
Collective unconscious. *See* Jung, Carl G.
Communist Manifesto, 118
Complexity (complex adaptive systems), 138–139. *See also* Chaos theory
Concrete operational stage. *See* Piaget, Jean
Confucianism, 21, 158, 160, 184–185
Connection, 29, 60, 67–68, 83, 93, 109, 115, 125, 159, 170, 175–177, 186
Consciousness. *See Nous*; Participation
Conviviality, 189
Crick, Francis, 52

Darwin, Charles, 53, 151–152
Dawkins, Richard, 75
Decentralization, 136, 138, 161–162, 182
Democracy. *See* Participation
Demoralization, 3, 5–7, 15–16. *See also* Rationalism
 political consequences of, 7–8, 18, 22
Descartes, René, 28, 31, 45–46, 155, 174
Diminishing returns, 62–65, 138
Discounting, 109
Domination, 2–3, 9, 28–30, 109
Dostoyevsky, Fyodor, 79, 92, 122
Durant, Will, and Ariel Durant, v, 16–17, 177
Dyson, Freeman J., 50

Eastern thought. *See* Confucianism; Needham, Joseph; Taoism
Ecological scarcity, ix–xi, 29–30, 64, 130, 164. *See also* Ecology; Thermodynamics
 and political transformation, 137–138, 188
Ecology, 25–43. *See also* Climax ecosystem; Ecological scarcity; Pioneer ecosystem; Systems
 and ecological man, 109–110
 and Gaia, 35–39, 126–127, 135, 163–164
 as model for political thought, 151–152, 159

moral vision of, 159, 196
and mutualism, 31, 35, 65
and symbiosis, 34–35, 176
Economic man, 109–110
Eddington, A. S., 49–50
Education. *See also Paideia*; Piaget, Jean
 in ecology and systems, 108–116
 and human ability, 115–116, 216n15, 221n33
 reform of, 108, 124–125
 as schooling, 98–99
Einstein, Albert, xii, 26, 49, 56, 68, 69, 78, 83, 106
Elite, 99–101, 116, 139. *See also* Natural aristocracy
Energy, diffuse versus concentrated, 65, 182
Energy slavery. *See* Slavery
Enlightenment
 philosophy and worldview of, 2–3, 9, 21–22, 170, 174–176, 195
 and progress, 81, 123, 158
 and religion, 5–6, 8, 84, 158
 and shift from Virgin to Dynamo, 159–160
Entropy, 36, 64. *See also* Thermodynamics
Epistemology. *See also* Metaphor; Noble lies; Rationalism
 and irrationality, 97
 and master metaphor, 43, 45–46, 78, 126, 133, 135
 and noble lies, 27, 119–120
 and relation to reality, 39, 49–50, 162
 twentieth century revolution in, 8, 21–22, 26, 46, 162, 176
Equality, 1–2, 4, 136–137, 147–148, 187–193
 in primal societies, 189–190
Eros, 91, 159–160, 170, 172, 178, 187
Ethics, 22, 26, 29, 33, 40, 43, 60, 93, 125. *See also* Morality; Mores; Natural law
Euclid, 46, 82
Euripides, 84
Evolution, 35–36, 53–54, 56, 79, 151–152
 of human mind, 85
 in humans, 70, 75
Exponential growth, 60–63, 163

False consciousness, 133–135
Farm subsides, 114, 154
Feynman, Richard P., 50
Fitzgerald, F. Scott, 76
Flannery, Tim, 178
Flows. *See* Income flows
Forests, 109–110, 141
Formal operations stage. *See* Piaget, Jean
France, Anatole, 191
Fraternity, 1–3, 136, 147–148, 165, 172, 186–193, 228n32
Freedom. *See* Liberty
French Revolution, 18, 143, 163
Freud, Sigmund, 28, 69–70, 80, 83–84, 88–89, 91
Froude, James Anthony, 1

Frugality, 138, 148, 180, 186–187, 190–191. *See also* Simplicity

Gaia. *See* Ecology
Galbraith, John Kenneth, 142
Gandhi, Mohandas, 126, 174
Gardner, Howard, 125
Gautama Buddha, 118
Geertz, Clifford, 179
Gleick, James, 56, 59
Goethe, Johann Wolfgang von, 77
Gould, Stephen Jay, 56
Government. *See also* Politics
  based on opinion, 133
  as necessary evil, 131
  role of, 129–133
  size and complexity of, 131, 136, 139, 150
Grand Inquisitor, 79, 92, 122–123
Guéhenno, Jean-Marie, 185

Haldane, J. B. S., 34
Hamilton, Alexander, 146, 149
Hegel, Georg Wilhelm Friedrich, 177, 191
Heisenberg, Werner, 49–50, 60
Hillman, James, 102
Hitler, Adolf, 134
Hobbes, Thomas, x, xiii, 7, 16–20, 79, 131, 152, 157, 161, 164, 183–84, 188
Homeostasis. *See* Balance
Hubris, 3, 20, 28–29, 59, 65, 67, 101

Human brain, 35, 51–52, 70, 75, 79, 94, 104
Human cognition. *See also* Human brain; Human nature; Metaphor
  forte of, 77
  limits of, 73–81
  as nonrational, 78–79
Human nature, 70–73, 81. *See also* Human cognition
  as archaic, 70, 86, 103
  and intrinsic morality, 82
  and passions, 17–19, 84, 92–93, 98, 131, 155
  as predatory, 171
Hume, David, 133
Humility, 29, 31–32, 60, 67–68, 83, 93, 109, 115, 125, 156, 159, 170, 186
Huxley, Aldous, 174

Ideas. *See* Plato
Ideology, 7–8, 80, 118–119
Illich, Ivan, 189, 192
Income flows, 65–68, 179
Individuation. *See* Jung, Carl G.
Indra's Net, 40
Instincts. *See* Jung, Carl G.
Interdependence. *See* Interrelationship
Interrelationship, 29, 33–35, 40, 60, 65, 67–68, 83, 93, 109
Involution, 179–181

James, William, 94
Jantsch, Erich, 55–56
Jaynes, Julian, 78
Jeans, James, 45, 47, 50, 83

Jefferson, Thomas, xii, 146–150, 164
  and agrarian way of life, 146–147
  and education, 149
  and natural aristocracy, 148–149
  and nondependence, 146
  and savage way of life, 147–148, 172, 174
  and setting of politics, 161, 183–184
  and ward-republics, 147, 149–150, 154–155, 184
Jung, Carl G., 69, 80, 83–93, 195
  and archetypes, 85–88, 93–94
  and collective unconscious, 69, 83, 85, 87, 89–90
  and Eros, 91 (see also Eros)
  and human destructiveness, 89–90
  and human weakness, 92–93
  and individuation, 89–93, 126, 174–178, 190–191
  and instincts, 71, 89–91, 104
  and mass psychology, 72 (see also Manias)
  and myth and religion, 84
  and necessity for elite, 100
  and objective psyche, 69, 83–85, 90
  and psyche as rhizome, 69, 83
  and spiritual cure for psychic ills, 89
  and technology, 227n18
  and *therapeia*, 94
  and tripartite division of humanity, 116–117
  and two million-year-old man, 70, 73, 86, 94, 97, 102–103, 121–122, 137, 165, 197
Justice. *See also* Plato
  and architecture, 153
  and ecology, 40–42

Kant, Immanuel, 82
Kazantzakis, Nikos, 187
Keats, John, 103
Keynes, John Maynard, xiii–xiv
Konner, Melvin, 71, 82, 105

Lao Tzu, 156, 174. *See also* Taoism
Lascaux, 169
Laws, man-made, 13–16, 18, 20, 102, 130, 136, 142, 191. *See also* Natural law
Le Bon, Gustave, 72, 80, 100, 150, 162
Lenin, Vladimir, 132
Lévi-Strauss, Claude, 73, 170
Liberation, 5, 195–197. *See also* Liberty
  from moral restraints, 196
  from nature, 28, 195–196
Liberty. *See also* Liberation; Locke, John
  of ancient gentes, 1–2, 192, 197
  and civil rights, 160, 191, 225n50
  in conflict with commonweal, 142
  and democracy, 150
  in liberal political philosophy, 39–40, 144–145, 188

Liberty (cont.)
 as license, 7, 16, 22
 in light of ecology, 40
 loss of, 4, 15–16, 18, 136
 in relation to equality and fraternity, 147–148, 187–193
Limits, xi, 29–31, 60, 62, 67–68, 93, 109
 as creative force, 31–32, 65, 112
Lippmann, Walter, 157
Locke, John, xiii, 19, 99, 144–145, 148, 184
Lorenz, Konrad, 84
Lovelock, James, 35, 65

Machiavelli, Niccolò, 183–184
MacLeish, Archibald, 134, 163
Madison, James, 148, 184
Manias, 5, 80–81, 87, 92
Margulis, Lynn, 35
Marx, Karl, 118, 133, 180
Master science, 43, 45–46, 78, 126, 135, 189
Material gratification, as rationale of political economy, 19–20
Maturana, Humberto, 37
McNeill, William, 184
McWilliams, Wilson Carey, 192
Meadows, Donella H., 113, 135
Mechanical worldview, 21–22, 26, 39, 45–47, 162, 164
Meritocracy, 99, 149, 190. *See also* Natural aristocracy

Metaphor, 47, 162. *See also* Epistemology; Master science; Noble lies; Story
 controlling power of, 133–135
 as critical for understanding, 77, 86, 105–106, 108
 as epoch shaping, 45–47, 97, 133, 162–163
 as forte of human mind, 78
 in Plato, 120–121
 in science, 27, 77–78
 as sign of genius, 78
Milgram, Stanley, 71
Mill, John Stuart, 130–131, 195
Millet system, 182
Mind. *See also* Human cognition; Human nature
 as inherent in matter, 45, 47, 49–50, 55, 60
 and *nous*, 37–39, 50, 54
Models. *See* Simulation
Moderation, 29, 33, 60, 67–68, 83, 93, 109, 115, 125, 159, 170, 186
Montaigne, Michel Eyquem de, 99
Montesquieu, Charles de Secondat, Baron de, 154–156
Moral entropy. *See* Morality
Morality, 125, 129. *See also* Ethics; Mores; Natural law
 in animals, 39
 basis of, 18, 25–26
 in human nature, 82
 immanent in nature, 8–9, 19, 21–22
 and moral entropy, 16–18, 81–82
More, Thomas, 174

Mores. *See also* Ethics;
　Morality; Natural law
　as basis of polity, 129,
　　144–145, 191
　as critical for good behavior,
　　13, 17, 19, 71, 82
　and religion, 157
Morgan, Lewis Henry, 1, 192
Murray, Charles, 115
Mutualism. *See* Ecology
Myth, 97, 121–127. *See also*
　Master science; Noble lies;
　Story
　function of, 79–80, 86–88
　and mythologems, 126
　and mythopoetic age, 175–176

Natural aristocracy, 99, 137,
　148–150, 190. *See also* Elite
Natural law, x, 22, 26–27, 93,
　165. *See also* Ethics;
　Morality; Mores
　contrast with man-made laws,
　　13–15, 41
　defined, 20–21
Needham, Joseph, 156
Neumann, Erich, 91–92
Newton, Isaac, 45–46, 78
Nietzsche, Friedrich, 25, 72
Noble lies (fictions), 27,
　116–124, 132–135. *See also*
　Metaphor; Myth; Plato; Story
Nous. *See* Mind

Objective psyche. *See* Jung,
　Carl G.
Oligarchy. *See* Elite
Optimization, 33, 67
Orwell, George, 164

Ottoman empire, 182
Overshoot and collapse, 55

Paideia, 82, 94, 97–127, 129,
　164. *See also* Education
　as aesthetic education,
　　104–108
　as Platonic liberal arts,
　　107–108
Paradigms. *See* Master science
Parmenides, 37
Participation
　as *participation mystique*, 1–2,
　　173–177
　in political life 132, 136, 147,
　　150
Pascal, Blaise, 104, 107–108,
　131
Passions. *See* Human nature
Pattern language. *See*
　Alexander, Christopher
Penrose, Roger, 106
Perception, 51–52, 73–77
Petroleum, 61–62, 66, 115,
　138
Philosophes. *See* Enlightenment
Physics, 45–68, 176
　as nonmechanical, 46–49
　as Platonic, 49–50, 52, 59
Piaget, Jean, 74, 115–116, 124,
　216n15
Picasso, Pablo, 169
Pioneer ecosystem, 31–32, 65.
　See also Climax ecosystem
Plato, xi, 115–123, 132–133,
　137, 145, 174, 192. *See also*
　Noble lies; *Paideia*; *Republic*
　and archetypes, 79, 85, 93–94
　and Confucius, 185

Plato (cont.)
 and human capacities, 116–117
 and Ideas, 50, 59, 85, 93–94
 and justice, 42, 153, 180–181
 and *Laws*, 222n37
 and liberal arts education, 108
 as mystic, 117
 and myth of Er, 122
 and physics, 47, 50, 60
 as poet of politics, 120–121
 and *Republic*, 117–119
Politeia, 82, 129–165
Politics. *See also* Government; Participation
 based on ecology, 151–152
 of consciousness, 93, 127, 134–136, 176–178, 196
 and economics of sufficiency, 137
 and effect of ecological scarcity, 137–138, 188
 and pattern language, 152–154
 in relation to setting, 143, 147–148, 155–156, 164–165
 and virtues of simplicity, 136–137
Pope, Alexander, xii, 131
Pragmatism
 in politics, 27
 in science, 222n40
Prigogine, Ilya, 53
Psychology, 69–94, 176. *See also* Freud, Sigmund; Jung, Carl G.
Psychosomatic disease, 87
Pythagoras, 50

Quality
 in ecosystems, 31–32, 65
 in education, 105, 107

Rank, Otto, 84
Rationalism, 6, 21–22, 25–27, 109, 124–125, 158. *See also* Epistemology; Reason; Systems
 as destructive, 97–98, 104, 176
 as irrational, 16–17, 21, 26–29, 80, 123, 155
 as opposed to instincts and passions, 84
 as tyrannical, 124
Reason, 5–6, 16, 21, 27, 80, 104, 131. *See also* Epistemology; Rationalism; Systems
 in Locke, 145
 in Montesquieu, 155
 and mythopoetic thought, 124
 and reasons of the heart, 104–108, 124–125
 reconstitution of, 108–115
Religion, 79
 different notions of, 157–158, 175–176
 ecology as basis for, 159
 in Freud, 84
 as necessary basis for polity, 157–161
Republic, xiii, 133, 174. *See also* Plato
 and difficulty of understanding, 117–119
 as political program, 118–119, 222n37

as thought experiment, 118–119
Robb, John, 161
Rousseau, Jean-Jacques, xi, 122, 146, 154–155, 164
  and civil faith, 157
  and freedom, 129, 144, 146, 155
  and general will, 129, 139–145
  and mores, 144, 172–173
  and natural aristocracy, 148
  and natural religion, 173–174
  and setting of politics, 161, 183–184
  and slavery to appetite, 129, 196
  and wiser savagery, 173–174
Rukeyser, Muriel, 78

Sade, Marquis de, 17–18, 157
Sahel, 111
Savage, 70, 73, 167–169
  aesthetic response of, 103–104
  and cognition, 76
  and concrete perception, 73–74
  as exemplar, 170
  and human nature, 169–170
  life of, 1–2, 73–75, 169–170, 176
  as predator, 171
Scarcity. *See* Ecological scarcity
Schiller, Friedrich, 106
Schrödinger, Erwin, 46, 49, 79
Schumacher, E. F., 189
Science, 134–135. *See also* Epistemology; Master science

Self-organization, 35–38, 52–55, 59, 156
Seneca, 184
Servan, J. M. A., 133
Shakespeare, William, 174
Shaw, Robert, 77
Shelley, Percy Bysshe, 127
Simeons, A. T. W., 87
Simplicity, 168. *See also* Frugality
  in politics, 136–139, 143, 147, 149–150, 153–156
Simulation, 57–58
Skinner, B. F., 145
Slater, Philip, 25
Slavery, 171
  to appetite, 129, 142, 192
  to energy, 171–172, 192
Smith, Adam, xiii, 19, 43, 145
Social change
  by osmosis, 162
  by upheaval, 163–164
Socrates, 42, 94, 118–123, 126, 164
Soleri, Paolo, 182–183
Sorokin, Pitirim A., 159
Stanford prison experiment, 71–72
Stocks. *See* Capital stocks
Stoicism, 157–158, 184–185
Story, 77–80, 86, 120, 122–123, 127, 175. *See also* Metaphor; Myth; Noble lies
Sublimation, 92
Sustainability, as oxymoron, xi, 29
Synergy in cultures, 180–181, 188

Systems, 109–115, 151–152, 176. *See also* Rationalism; Reason
Systems operations stage, 116, 124. *See also* Piaget, Jean
Symbiosis. *See* Ecology

Tacitus, Cornelius, 13–14
Taine, Hippolyte, 13, 132
Tainter, Joseph A., 138
Taoism, 40, 156–159. *See also* Lao Tzu
Technology, 29–30, 162–163. *See also* Appropriate technology
Thales of Miletus, 186, 189
Therapeia, 82, 92, 94, 97, 103, 129
Thermodynamic costs. *See also* Entropy; Thermodynamics
  internalization of, 31, 65, 67
  as tax levied on human transactions, 63–64
Thermodynamics. *See also* Entropy; Thermodynamic costs
  laws of, xi, 63–65, 151, 162, 182
  in relation to money economy, 67
Thomas, Lewis, 37, 105
Thoreau, Henry David, 107
  and politics, 183–184
  and simplicity, 167–168, 174, 187
  and technology, 227n18
  and wiser savagery, 167–168, 193

Timeless way. *See* Alexander, Christopher
Tocqueville, Alexis de, 187–188
Tsetse fly, 112
Two million-year-old man. *See* Jung, Carl G.

Utopia, 5, 98, 137, 174, 188

Virtue. *See also* Ethics; Morality; Mores
  as basis of politics, 16–19, 22, 72–73, 82
  demotion of by Hobbes, x, 7
  and natural aristocracy, 148–149
  and noble life, 122, 138
  as self-perfection, 119
  and wisdom, 94, 98, 101, 120, 129, 136, 156
Voltaire, 3, 157

Waldrop, M. Mitchell, 53–54
Water, 66–67
Weber, Max, 170
Weinberg, Steven, 103
Western Inscription, 161
Whitehead, Alfred North, 68, 106
Wittgenstein, Ludwig, 52, 79–80

Yeats, William Butler, 125

Zeno, 25
Zhang Zai, 160
Zimbardo, Philip, 71–72